Also published in this series

:oncept of Evidence, edited by Peter Achinstein
:eptual Knowledge, edited by Jonathan Dancy
Moral Concepts, edited by Joel Feinberg
Theories of Ethics, edited by Philippa Foot
ilosophy of History, edited by Patrick Gardiner
Philosophy of Mind, edited by Jonathan Glover
ledge and Belief, edited by A. Phillips Griffiths
cientific Revolutions, edited by Ian Hacking
d Economic Theory, edited by Frank Hahn and Martin Hollis
Commands and Morality, edited by Paul Helm
Hegel, edited by Michael Inwood
ilosophy of Linguistics, edited by Jerrold J. Katz
rence and Modality, edited by Leonard Linsky
Philosophy of Religion, edited by Basil Mitchell
Concept of God, edited by Thomas V. Morris
Priori Knowledge, edited by Paul K. Moser
Aesthetics, edited by Harold Osborne
heory of Meaning, edited by G. H. R. Parkinson
Philosophy of Education, edited by R. S. Peters
tical Philosophy, edited by Anthony Quinton
Practical Reasoning, edited by Joseph Raz
osophy of Social Explanation, edited by Alan Ryan
entialism and its Critics, edited by Samuel Scheffler
Philosophy of Language, edited by J. R. Searle
nantic Syntax, edited by Pieter A. M. Seuren
Applied Ethics, edited by Peter Singer
hilosophical Logic, edited by P. F. Strawson
on Human Understanding, edited by I. C. Tipton
eories of Rights, edited by Jeremy Waldron
on Pure Reason, edited by Ralph C. S. Walker
Free Will, edited by Gary Watson
Philosophy of Action, edited by Alan R. White
Metaphysics and Philosophy of Science, edited by R. S. Woolhouse

Other volumes are in preparation

OXFORD RE

PROPOSITIC

The
Pe

The
The
Knc

Philosophy

Divi

The
Re
The
Th

The
The
P

The Ph
Conse
Th

Loc

Ka

Th
Leibn

PROPOSITIONS AND ATTITUDES

EDITED BY

NATHAN SALMON

AND

SCOTT SOAMES

OXFORD UNIVERSITY PRESS

1988

Oxford University Press, Walton Street, Oxford OX2 6DP
Oxford New York Toronto
Delhi Bombay Calcutta Madras Karachi
Petaling Jaya Singapore Hong Kong Tokyo
Nariobi Dar es Salaam Cape Town
Melbourne Auckland
and associated companies in
Berlin Ibadan

Oxford is a trade mark of Oxford University Press

Published in the United States
by Oxford University Press, New York

British Library Cataloguing in Publication Data
Propositions and attitudes.—(Oxford readings in philosophy).
1. Logic. Propositions
I. Salmon, Nathan U. II. Soames, Scott
160
ISBN 0–19–875092–7
ISBN 0–19–875091–9 (Pbk.)

Library of Congress Cataloging in Publication
Propositions and attitudes.
(Oxford readings in philosophy)
Bibliography: p. Includes index.
1. Proposition (Logic) 2. Belief and doubt.
3. Knowledge, Theory of. I. Salmon, Nathan U.
1951– . II. Soames, Scott. III. Series.
BC181.P75 1988 160 88–9862
ISBN 0–19–875092–7
ISBN 0–19–875091–9 (Pbk.)

Printed in the United States of America

CONTENTS

ACKNOWLEDGEMENTS

1. Bertrand Russell: 'Knowledge by Acquaintance and Knowledge by Description', *Proc. of the Aristotelian Society* 11. © The Aristotelian Society, (1910–1911). Reprinted by courtesy of The Editor.

2. Gottlob Frege: 'Thoughts' from *Collected Papers On Mathematics, Logic, and Philosophy*, ed. Brian McGuinness (1984). Reprinted by Permission of Basil Blackwell Ltd.

3. Selection from the Frege-Russell correspondence, 13 November 1904 and 12 December 1904, from Gottlob Frege, *Philosophical and Mathematical Correspondence*, translated from the German by Hans Kaal, abridged for the English edition by Brian McGuinness, © 1980. Reprinted by Permission of Basil Blackwell Ltd., and the University of Chicago Press.

4. Alonzo Church: 'A Remark Concerning Quine's Paradox About Modality'. Reprinted by Permission of the author. Spanish version published in *Analisis Filosófico* 2, 1982.

5. David Kaplan: 'On The Logic of Demonstratives', *Journal of Philosophical Logic* 8, (1979), pp. 81–98. All rights reserved. Copyright © 1979 by David Kaplan. Reprinted by permission of the author.

6. John Perry: 'The Problem of the Essential Indexical', *Noûs*, 13, 1979, pp. 3–21. Reprinted by permission of the author.

7. Saul Kripke: 'A Puzzle about Belief', *Meaning and Use* edited by A. Margalit (Dordrecht, D. Reidel), pp. 239–83. Copyright © 1979 by Saul A. Kripke. Reprinted by permission of the author.

8. Hilary Putnam: 'Synonymy, and the Analysis of Belief Sentences', *Analysis* 14, 1954, pp. 114–22. Reprinted by permission of the author.

9. Alonzo Church: 'Intensional Isomorphism and Identity of Belief', *Philosophical Studies* 5, 1954 pp. 65–73. Reprinted by permission of the author.

10. Mark Richard: 'Direct Reference and Ascriptions of Belief', *Journal of Philosophical Logic* 12 (1983), pp. 425–52. Copyright © 1983 by D. Reidel Publishing Company. Reprinted with permission.

11. Scott Soames: 'Direct Reference, Propositional Attitudes, and Semantic Content' originally published in *Philosophical Topics* 15, 1987, pp. 47–87. © 1987 The Board of Trustees of the University of Arkansas. Used with permission.
12. Nathan Salmon: 'Reflexivity', originally published in the *Notre Dame Journal of Formal Logic* 27, Number 3 (July 1986) pp. 401–29. Copyright © 1986 Nathan Salmon. Reprinted by permission of the author.

INTRODUCTION

NATHAN SALMON
SCOTT SOAMES

THE concept of a *proposition* is important in several areas of philosophy, and central to the philosophy of language. What is a proposition? Some features of propositions seem to be essential to the very concept: If you utter the words 'Snow is white' and a French speaker utters the words 'La neige est blanche', there is some sense in which both of you say the same thing despite your having used different words. This thing that both of you said is a proposition: the proposition that snow is white. When uttering or writing a declarative sentence (in a given context) one asserts (or records) a piece of information, which is the semantic information content of the sentence (in the context). Since they are the contents of declarative sentences—and what one asserts in uttering declarative sentences—propositions are the sorts of things that are true or false. But making true or false assertions is not the only thing we do with propositions. We also bear cognitive attitudes toward them. Propositions are what we believe, disbelieve, or suspend judgement about. When you fear that you will fail or hope that you will succeed, when you venture a guess or feel certain about something, the object of your attitude is a proposition. That is what propositions are.

The readings collected in this volume investigate many different philosophical issues concerning the nature of propositions and the attitudes we bear to them. Although each of the readings stands on its own as a self-contained unit, there are a number of inter-connections between them. In the remainder of this Introduction we shall present one line of thought connecting them all. While not attempting to deal with all of the significant issues addressed in individual selections, this line of thought illustrates one way of approaching the collection as a whole.

It has often been observed that the content of a sentence must be determined in some manner by its grammatical constituents and the roles they play in the sentence—otherwise we could not understand what information is expressed in new sentences seen or heard for the first time. It is apparent that each grammatical constituent of a sentence (or at least each typical one) makes some contribution of its own to securing the content of the sentence. We may call the contribution made by an expression its 'content'. The content of a sentence is thus a function of the separate contents of its grammatical constituents and the manner in which they are combined in constructing the sentence, in the sense that the following principle of Compositionality obtains:[1]

If S and S' are sentences that are grammatically constructed in the same way from corresponding constituents having the same content, then S and S' have the same content.

This semantic principle accounts for the fact that the English words 'Snow is white' and their word-for-word French translation 'La neige est blanche' form grammatical units having the same content. The principle also has the important consequence that any two expressions of a single language that share the same content (for example, 'lawyer' and 'attorney') are interchangeable within a sentence without altering its content, and hence without altering its truth value.

Some expressions in natural language have the special property that (when they are used in the normal way) they stand for, or *refer to* (denote, designate), some particular person, place, or thing—like the expressions 'Bertrand Russell' (proper name), 'he' (demonstrative pronoun), and 'the author of *Waverley*' (definite description). One idea that naturally springs to mind is that the content of a referring expression is its referent (what it refers to). This has often been called 'the naïve theory'. This theory is especially compelling in the case of at least one sort of referring expression: free individual variables occurring in an open sentence—for example, the occurrence of 'x' in 'x is pretty'. Once the city of London is assigned as the referent (value) of 'x', there is nothing else for the variable to contribute to the content of 'x is pretty', under this assignment. The naïve theory is also especially compelling in the case of certain uses of the pronouns 'he', 'she', and 'it'—which seem to function in some contexts in the same way as variables of formal logic. The naïve theory simply extends these observations concerning variables and pronouns to all referring expressions.[2]

It is well-known, however, that this extension, in combination with Compositionality, leads to puzzles and difficulties involving substitutions of certain coreferential expressions, especially proper names and definite descriptions. One of the most famous of these puzzles was introduced by Gottlob Frege and discussed independently by Bertrand Russell. We shall consider Russell's example:

(1) Scott is the author of *Waverley*.

The content of this sentence is a proposition that King George IV was at one time not certain about, since (as a matter of historical fact) he was not certain whether Scott wrote *Waverley*. Nevertheless (1) is true. Therefore, the expressions 'Scott' and 'the author of *Waverley*' have the same referent. This conflicts with the naïve theory. For according to that theory in conjunction with Compositionality, the name 'Scott' may be substituted for the definite description 'the author of *Waverley*' in (1) without altering its content. Thus (1) should have the same content as

(2) Scott is Scott

But King George was, at the time in question, fully certain that Scott is Scott. Since King George was certain about this without being certain that Scott is the author of *Waverley*, the sentences (1) and (2) differ in content, contrary to the naïve theory.

In his classic article 'On Denoting' (*Mind* 1905), Russell presented an elegant solution to this substitutivity puzzle. He claimed that, despite appearances, sentence (1) is not really an identity statement and the phrase 'the author of *Waverley*' is not really a referring expression. According to Russell, the semantic function of a description—whether indefinite ('some author', 'an author', 'no author') or definite ('the author of *Waverley*')—is not to contribute a constituent to propositions at all. Instead, sentences containing descriptions are convenient abbreviations for more complex sentences lacking them.

For example, sentence (1) is treated as an abbreviation of

(3) There is an individual x such that x and one else wrote *Waverley*, and Scott $= x$.

This sentence is true if and only if for some value of the variable 'x' the proposition expressed by the following is true:

(4) x and no one else wrote *Waverley*, and Scott $= x$.

Thus, according to Russell, (1) expresses (a proposition trivially equivalent to) the proposition directly concerning Scott that he and no one else wrote *Waverley*. Although Scott is a constituent of this proposition, he is contributed by the name 'Scott' rather than by the definite description 'the author of *Waverley*', which has been "broken up" in (3). Since these expressions do not make the same contribution to the propositions expressed by sentences in which they occur, substitution of one for the other does not preserve the proposition expressed.

Russell distinguished descriptions ("denoting phrases") from what he called 'logically proper names' or 'names in the strict logical sense'. A *logically proper name* is a genuine referring expression whose contribution to the propositions expressed by sentences containing it is precisely its referent. (In more recent terminology, due to David Kaplan, such expressions are often called 'directly referential'.) In illustrating Russell's theory of descriptions above, we treated the name 'Scott' and the variable 'x' in (4) as logically proper names. Russell's theory analyses the definite-description operator (or definite article) 'the' as eliminable in favour of quantifiers, connectives, and the identity predicate—thereby eliminating definite descriptions. Variables remain as logically proper names, in the sense that, under any assignment of a value to a variable, the variable serves as a logically proper name of its assigned value.

In his essay 'A Remark Concerning Quine's Paradox About Modality' (see Chapter 4), Alonzo Church presents a strengthened substitutivity puzzle in which definite descriptions are replaced by variables. This twist on the Frege–Russell substitutivity puzzle yields an interesting paradox, to which Russell's theory of descriptions is not directly applicable. Church's paradox concerns belief. Church notes that it is at least very likely that, throughout his lifetime, George IV did not believe anyone to be not identical with himself. That is, for every value of the 'x' the following is true (with respect to any time):

(5) George IV does not believe that $x \neq x$.

We are supposing that variables are logically proper names. By Compositionality, then, the following is also true whenever the variables 'x' and 'y' are assigned the same value:

(6) George IV does not believe that $x \neq y$.

Therefore, for all values of 'x' and 'y' (whether they are the same or not) the following is true:

(7) If George IV believes that $x \neq y$, then $x \neq y$.

But this seems to attribute to King George the incredible power to make individuals distinct merely by believing them distinct. Such attribution seems to be falsified in Russell's example if we imagine that, at some appropriate time, King George believed of the author of *Waverley* and Scott that the former is not the latter. By a simple extension of Church's paradoxical derivation one may derive that King George has an even more majestic power: the power to make individuals distinct by proclamation. For if King George never says of anything that it is not identical with itself, then by an argument exactly analogous to Church's, for every value of 'x' and 'y' the following is true:

(8) If George IV says that $x \neq y$, then $x \neq y$.

Surely, no one has such power.

The problem can be illustrated using a slight modification of Russell's example. Let us suppose that at a book-signing ceremony given by "the author of *Waverley*", a cleverly disguised Scott autographs King George's copy of *Waverley*. King George, being fooled by Scott's disguise, concludes that *Waverley* was written by someone other than Scott. He sincerely declares,

(9) He is not Scott,

pointing to the disguised author. According to the naïve theory, King George's utterance has the same content as

(10) Scott is not Scott.

Yet King George surely disbelieves, and would vigorously deny, that Scott is not himself. Because of this, Church would maintain that sentences (9) and (10) differ in content every bit as much as sentences (1) and (2). But Church's argument makes the situation considerably worse than before. For if (7) and (8) are true, then it would seem that King George would have to be right when he utters (9). Moreover, Russell's theory of descriptions cannot be directly applied (as it was in the case of (1) and (2)) to solve Church's paradox.[3]

In constructing this paradox we assumed that the demonstrative pronoun 'he' and the ordinary proper name 'Scott', as used by King George, are logically proper names (in Russell's sense), and hence that the proposition expressed by King George's assertive utterance of (9) is the same as the proposition expressed by '$x \neq y$' when 'x' and 'y' are both assigned Scott as their value. One possible response to the paradox is to reject this assumption in favour of the view that one or both of the terms in (9) are disguised definite descriptions. This is the view taken by Russell in 'Knowledge by Acquaintance and Knowledge by Description' (see Chapter 1).[4] There he maintains the principle that in order to apprehend a proposition (and hence in order to believe it) one must bear a very special epistemic relation, which he called 'acquaintance', to each of the proposition's constituents. According to Russell, the only things we can bear this relation to are things whose identity and distinctness we cannot be mistaken about—things like our own sense experiences, as well as abstract objects like properties and relations. Given this, Russell's "principle of acquaintance" precludes the possibility of King George's believing any proposition in which Scott occurs directly as a constituent. Moreover, if the values of 'x' and 'y' in (7) are things George is acquainted with, Russell would insist that they are distinct whenever they are believed distinct, in which case (7) is indeed true. On the other hand, if one or both are things King George is not acquainted with, then Russell would insist that (6) is automatically true (since King George cannot apprehend any proposition directly concerning such things); in which case (7) is true by virtue of having a false antecedent, and hence is harmless.

Although Russell's principle of acquaintance appears to provide a possible solution to Church's paradox about belief, it does so at the heavy cost of severely restricting what one can believe. The principle renders propositions directly concerning the objects around us (many of which are of vital importance in our lives) cognitively inaccessible; they cannot be believed, disbelieved, or even apprehended. Worse, the principle fails to resolve all versions of the paradox. Even if King George cannot make a mistake about the identity of things he is acquainted with, he can knowingly deny true identity statements concerning them. Thus, a version of the paradox involving (8) remains. If King George never says of any object of his acquaintance that it is distinct from itself, then his saying of any such objects that they are distinct guarantees that they are distinct.

In fact, the paradox appears to remain in the case of assertions concerning other objects as well. When King George utters (9), does he not make an assertion directly concerning Scott? Suppose a friend of yours says to you

(11) You are a native Californian.

Russell would claim that the word 'you' in (11) abbreviates some definite description, such as 'the person I am addressing'. On Russell's analysis, then, what your friend asserted in uttering (11) is that there is an individual *x* such that he is addressing *x* and only *x*, and *x* is a native Californian. But, as David Kaplan has argued, this cannot be right. What your friend said is a proposition that would have been true even if he had never existed and had never addressed anyone, as long as you had been born in California.

Kaplan (see Chapter 5) emphasizes that sentences containing indexicals—context-sensitive expressions like 'you', 'I', and 'he'—express different propositions in different contexts of use. The proposition expressed in a given context may be true in some circumstances and false in others. For example, the proposition expressed by (11), when addressed to a person *A* in a given context, is true in a given circumstance if and ony if *A* is born in California in that circumstance. On Kaplan's view, this proposition directly concerns *A*, since 'you' functions as a logically proper name. Likewise, on Kaplan's view, the indexical words 'I', 'he', 'she', 'this', and 'that', when used in the ordinary way, are logically proper names (i.e. directly referential).[5]

Kaplan points out that indexical sentences having different meanings can nevertheless express the same proposition when uttered in different but suitably related contexts. For example, you can assert what your friend asserted when he uttered (11) by saying.

(12) I am a native Californian.

Although these indexical sentences differ in meaning, the proposition expressed by (11) in your friend's context is the same as the proposition expressed by (12) in yours. Facts like these indicate the need to distinguish the meanings of sentences from the propositions they express. The meaning of an expression (sentence) may be thought of as a rule that specifies a content (proposition) for any possible context. Kaplan calls this rule the 'character' of the expression.

John Perry applies this account to issues involving belief in his contribution to this volume (see Chapter 6). By analogy to Kaplan's distinction between the content and the character of a sentence, Perry draws a distinction between the proposition that someone believes and his *belief state* in virtue of which he believes this proposition. Perry holds that when your friend sincerely utters (11) while addressing you, and you sincerely utter (12), you and your friend believe the same proposition in virtue of being in different belief states. Similarly, everyone who sincerely utters (12) will be in the same belief state, although each person in this state believes a different proposition.

On Kaplan's account of indexicals, King George says something directly concerning Scott when he utters (9), by virtue of his use of the demonstrative pronoun 'he'. According to Kaplan and Perry, King George also believes this proposition directly concerning Scott. Both versions of Church's paradox thus remain on this account—provided that the name 'Scott' is also a logically proper name.

Are ordinary proper names logically proper, or are they descriptive in content as Frege and Russell claimed? The latter view has been subjected to devastating criticism, most notably by Saul Kripke in *Naming and Necessity*.[6] In his contribution to this collection (see Chapter 7), Kripke challenges objections to the former view that are based on substitutivity problems of the sort generated by (9) and (10).[7] In the course of doing this, Kripke demonstrates that difficulties involving the use of ordinary proper names in specifying someone's beliefs arise whether one treats such names as logically proper, views them as essentially descriptive, or makes no explicit decision at all about their content. Indeed he shows that such difficulties can be generated independently of standard substitutivity principles. Kripke concludes that one should not draw any morals about the contents of names or the truth of substitutivity principles from the existence of substitutivity puzzles.

Kripke's examples suggest that the source of our substitutivity puzzles does not lie in the claim that coreferential proper names, indexicals (relative to contexts), and variables (relative to assignments of values) have the same content. Further support for this suggestion is provided by the fact that analogous substitutivity puzzles arise whenever there are different expressions having the same content. Suppose that the words 'lawyer' and 'attorney' are exact synonyms, and hence have the same content. The conjunction of this assumption with Compositionality yields the result that the sentences 'Lawyers are lawyers' and 'Attorneys are lawyers' express the same proposition. Many, however, question this conclusion, on the grounds that it is possible to assert and believe the former proposition while denying and disbelieving the latter.

In 1950 Benson Mates gave this problem an interesting twist. Consider the following sentences:

(13) Whoever believes (asserts) that lawyers are wealthy believes (asserts) that lawyers are wealthy.

(14) Whoever believes (asserts) that lawyers are wealthy believes (asserts) that attorneys are wealthy.

The truth of (13) is beyond doubt. What about (14)? Imagine that Vladimir is under the misimpression that 'attorney' means *law student*. Vladimir believes that all lawyers are wealthy, but that not all law students are. Does he believe that attorneys are wealthy? Many theorists, such as Church, would insist that he does. Others disagree.[8] They agree that Vladimir believes that lawyers are wealthy, but deny that he believes that attorneys are. It would seem, then, that they doubt the content of (14) without doubting that of (13), and that sentences (13) and (14) therefore express different propositions. But how can they, since they differ only in the substitution of one synonym for another?

In his classic 1954 article on identity of belief (see Chapter 9) Church uses standard practices of translation to give a powerful argument that (13) and (14) cannot differ in content (in English). (These translation practices play a similar role in the central argument in Kripke's contribution.) Hilary Putnam makes an alternative proposal in his contribution (Chapter 8). According to him, (13) and (14) have different contents—even though they are grammatically constructed in the same way from corresponding constituents with the same content. The difference in the contents of (13) and (14) is attributed to a difference in logical structure, arising from the fact that (13) contains two occurrences of a single constituent, 'lawyer', where (14) contains occurrences of different constituents. Putnam appeals to this notion of logical structure in rejecting the original Compositionality principle formulated above, and replacing it with the following more restrictive principle:[9]

If S and S' are sentences that have the same logical structure and their corresponding constituents have the same content, then S and S' have the same content.

The effect of this replacement is not limited to examples of the sort Putnam had in mind, like (13) and (14), but extends to the original puzzle involving (1) and (2), and to all of the versions of Church's paradox that we have discussed.[10] In particular, it allows one to maintain that (5) is true for every value of 'x', even though (6) is false when Scott is the value of both 'x' and 'y'. Putnam's proposal thus blocks Church's derivation of the paradoxical (7) from (5) by restricting substitutions to those that preserve logical structure. As a result, Putnam's proposal allows one to maintain the thesis that, if x and y are the same individual, someone can deny and disbelieve that $x = y$ without thereby denying or disbelieving that $x = x$.

This thesis raises a difficult question that we have been holding in abeyance and that we must now address: What does King George assert, and believe, when he utters (9)? If his utterance does not have the same content as (10), then what exactly does he assert (believe)? Is it obvious that he does not assert (believe) that Scott is not Scott?

A number of theorists would insist (or are otherwise committed to the claim) that even if the expressions 'he' and 'Scott' in his utterance of (9) are logically proper names, King George still does not have the contradictory belief that Scott is not Scott. For if he had that belief, it would seem that he should also believe that someone is not himself. Surely King George does not have this belief; no one does. This is the position taken by Mark Richard.[11] In his contribution to this collection (see Chapter 10) Richard develops an account of belief ascriptions containing logically proper names that is similar in outline to Putnam's treatment of (13) and (14) above.[12]

By contrast, we maintain that when King George sincerely utters (9), he does in fact assert and believe that Scott is not Scott—even though he would not assent to (10). Thus, we hold that (5) is false, when Scott is assigned as the value of 'x'. An analysis of belief (and assertion) ascriptions leading to this result is given by Soames in his contributed article (see Chapter 11)—an account whose treatment of ascriptions containing logically proper names is broadly similar to Church's treatment of (13) and (14). Soames insists that although King George asserts and believes that

Scott is not Scott, he does not assert or believe that someone is distinct from himself. Soames uses this distinction between what King George asserts (believes) and what he does not assert (believe) to undermine a popular conception of semantic content according to which the content (i.e. proposition) expressed by a sentence in a context is something like the set of circumstances with respect to which the sentence, as used in the context, is true. Soames proposes to replace this conception with one in which classical Russellian propositions are semantic contents of sentences. These Russellian propositions determine, but are not determined by, the circumstances in which the sentences are true. This conception of the contents of sentences agrees with Russell regarding the objects of propositional attitudes, while departing from him in allowing that we frequently assert and believe propositions directly concerning individuals with whom we are not acquainted, in Russell's special sense.

In his contributed article (see Chapter 12), Salmon attempts to take the sting out of our claim that when King George sincerely utters (9), what he asserts and believes is precisely that Scott is not Scott, and that (5) is therefore false under the assignment of Scott to '*x*'. Drawing a distinction between the proposition that Scott is not Scott and the proposition that Scott is not self-identical, Salmon argues that although King George asserts and believes the former in uttering (9), he does not thereby assert or believe the latter. In fact, Salmon argues, it is very likely that for every *x*, King George does not assert or believe that *x* is not self-identical—a fact that can easily mislead one into thinking that (5) is always true.

If King George believes that Scott is not Scott, as we maintain, why then does he not assent to (10) instead of vigorously dissenting from it? In answering this question we appeal to a distinction similar to Perry's between belief states and their contents in a context. In particular, Salmon argues that although King George fully understands both (9) and (10), when he hears or sees (10) he recognizes its content as a logical impossibility, and rejects it, but when he sincerely utters (and understands) (9) he takes its content in a completely different way, failing to recognize it as the very same contradictory proposition that is expressed by (10). Thus, his sincere utterance of (9) and his rejection of (10) indicate, respectively, his belief and his disbelief of one and the same proposition.

Notes

1. A more discriminating formulation of Compositionality is needed for languages whose sentences involve devices like quotation marks, which induce expressions occurring in their scope not to contribute what they customarily contribute to the contents of sentences. The general intent of the principle is clear enough for the present purpose, since we shall not be concerned in this Introduction with sentences involving quotation marks or similar devices.
2. For further elaboration of the naïve theory (and some appropriate modifications), see Nathan Salmon, 'Reference and Information Content: Names and Descriptions', in D. Gabbay and F. Guenthner (eds.), *Handbook of Philosophical Logic, iv. Topics in the Philosophy of Language* (Dordrecht: D. Reidel, forthcoming); and his *Frege's Puzzle* (MIT Press/Bradford Books, 1986), 16–43. In the excerpt included in this volume from his 1904 correspondence with Bertrand Russell (ch. 3), Gottlob Frege rejects the view that the content of a proper name is simply its referent. In his contributed article 'Thoughts' (ch. 2), he offers a rival account of the contents of sentences containing proper names or other referring expressions.
3. Church himself suggests that the paradox provides a strong reason for rejecting treatments of quantifying-in which take variables to be logically proper names and which preserve the standard laws of logic.
4. In this paper, published in 1911, Russell adopts the view that ordinary proper names and demonstrative pronouns, when used to refer to other persons or external objects, are disguised definite descriptions. This view runs counter to the main drift of the excerpt from Russell's 1904 correspondence with Frege.
5. See 'Demonstratives' in J. Almog, J. Perry, and H. Wettstein (eds.), *Themes from Kaplan* (Oxford University Press, 1988). This view of indexicals contrasts with that given by Frege in 'Thoughts'. For a criticism of Frege's view see John Perry, 'Frege on Demonstratives', *The Philosophical Review* 86 (1977), 474–97.
6. Harvard University Press and Basil Blackwell, 1980 (originally published 1972). See also his 'Identity and Necessity', in M. Munitz (ed.), *Identity and Individuation* (New York University Press, 1971), 135–64, also in S. Schwartz (ed.), *Naming, Necessity, and Natural Kinds* (Cornell University Press, 1977), 66–101; Keith Donnellan, 'Proper Names and Identifying Descriptions', in D. Davidson and G. Harman (eds.), *Semantics of Natural Language* (Dordrecht: D. Reidel, 1972), 356–79, and 'Speaking of Nothing', *The Philosophical Review* 83 (1974), 3–32, also in Schwartz, pp. 216–44; Frederic Fitch, 'The Problem of The Morning Star and the Evening Star', *Philosophy of Science* 16 (1949), 137–41; David Kaplan 'Demonstratives', in J. Almog, J. Perry, and H. Wettstein (eds.), *Themes from Kaplan* (1988); Ruth Barcan Marcus, 'Modalities and Intensional Languages', *Synthese* 13 (1961), 303–22, and 'Discussion of the Paper by Ruth B. Marcus' (with Quine, Kripke, McCarthy, and Follesdal), *Synthese* 14 (1962), 132–43; Hilary Putnam, 'Is Semantics Possible?' in H. E. Kiefer and M. Munitz (eds.), *Language, Belief, and Metaphysics* (State University of New York Press, 1970), 50–63, also in Schwartz, pp. 102–118, 'Meaning and Reference', *Journal of Philosophy* 70 (1973), 699–711, also in Schwartz, pp. 119–132, and 'The Meaning of "Meaning"', in K. Gunderson (ed.), *Language, Mind, and Knowledge* (University of Minnesota Press, 1975), 131–93, also in Putnam's *Philosophical Papers, ii. Mind, Language and Reality*

(Cambridge University Press, 1975), 215–71; Nathan Salmon, *Reference and Essence* (Princeton University Press and Basil Blackwell, 1981), Part I; and Arthur Smullyan, Review of Quine's 'The Problem of Interpreting Modal Logic', *The Journal of Symbolic Logic* 12 (1947), 139–41.

7. Salmon, *Frege's Puzzle*, provides further arguments along these lines, as well as arguments favouring and elaborating the doctrine that ordinary proper names (and other simple referring expressions) are logically proper names.
8. See e.g. Tyler Burge, 'Belief and Synonymy', *Journal of Philosophy* 75 (1978), 119–38.
9. Church's translation argument (see ch. 9) was intended, at least in part, as an objection to this proposal of Putnam's.
10. Putnam's proposal, however, does not provide solutions to all Mates-type puzzles. Suppose Mary fully understands both 'lawyer' and 'attorney' (recognizing them as exact synonyms), and expresses her conclusions concerning Vladimir's doxastic state by saying "Vladimir believes that all lawyers are wealthy but he does not believe that all attorneys are wealthy. On the contrary, he believes that not all attorneys are wealthy". It would seem, then, that the sentences 'Vladimir believes that lawyers are wealthy' and 'Vladimir believes that attorneys are wealthy' express different propositions, since Mary believes one and disbelieves the other. Both the original Compositionality principle and Putnam's substitute preclude this. Instead, Putnam would likely claim that Mary believes that Vladimir believes lawyers are wealthy, and conclude from this that she also believes that Vladimir believes attorneys are wealthy, since on Putnam's proposal, these propositions concerning what Vladimir believes are the very same.

    Nevertheless, according to Putnam, Mary disbelieves that if Vladimir believes lawyers are wealthy, then he believes attorneys are wealthy. He would likely claim, therefore, that she also disbelieves the stronger, conjunctive proposition that: Vladimir believes lawyers are wealthy, and he believes attorneys are wealthy. Thus Putnam attributes to Mary the apparently illogical position of believing a proposition (that Vladimir believes lawyers, i.e. attorneys, are wealthy) while disbelieving the conjunction formed from this proposition and itself, and even disbelieving the conditional formed from this proposition together with itself.

    On the basis of Mary's remarks Putnam would likely claim further that Mary believes that Vladimir does not believe attorneys are wealthy, and that she also believes (the conjunctive proposition) that: Vladimir believes lawyers are wealthy and he does not believe attorneys are wealthy. Thus on Putnam's account, Mary believes both a proposition and its negation, and even believes the conjunction of this proposition and its negation. In light of these apparently harsh verdicts about Mary, the fact that Putnam's proposal postulates a distinction between the content of (13) and that of (14) no longer seems a significant improvement over original Compositionality. Even the original principle allows that Mary may (inconsistently) believe the content of (13) while disbelieving that of (14).

    For further criticism of Putnam's proposal, see Israel Scheffler, 'On Synonymy and Indirect Discourse', *Philosophy of Science* 22 (1955), 39–44, at p. 42 n. 7; Salmon, *Frege's Puzzle*, pp. 164–5, n. 4. See also Scott Soames, 'Substitutivity', in J. J. Thomson (ed.), *On Being and Saying: Essays for Richard Cartwright* (MIT Press, 1987), where Putnam's proposal, and extensions of it to deal with standard substitutivity puzzles involving singular terms, are discussed and criticized at length.

By contrast to Putnam, Church would insist that Mary ('whatever she herself may tell us') does not doubt that if Vladimir believes lawyers are wealthy, then he believes attorneys are wealthy. Church would likely claim further that Mary does not doubt that Vladimir believes attorneys are wealthy. What Mary really doubts, Church maintains, is something linguistic, for example, that the sentence 'Vladimir believes that attorneys are wealthy' is true in English. (In the final footnote to his contributed article, Kripke expresses sympathy for something like Church's position.)

This conclusion, however, is questionable. Suppose that Mary is trained in semantics, and knows to distinguish the content of any English sentence *S* from the separate proposition that *S* is true in English. Nevertheless, she maintains her conclusions concerning Vladimir. In particular, she correctly understands the sentence 'Vladimir believes that attorneys are wealthy' to express (in English) that Vladimir believes attorneys, i.e. lawyers, are wealthy (rather than something linguistic); but she sincerely dissents from this sentence (as a sentence of English) because of her (mistaken) philosophical views about the objects of belief or the truth conditions of belief ascriptions. (Cf. n. 8, above. Burge may be in a similar state.) Under these circumstances, Mary's sincere dissent seems to indicate a doubt (however confused) that Vladimir believes attorneys (i.e. lawyers) are wealthy—in addition to her linguistic doubt that the sentence 'Vladimir believes that attorneys are wealthy' is true in English, and in addition to her correct belief that Vladimir believes lawyers are wealthy. See Section IX of Soames, 'Substitutivity', for further criticism of Church along these lines. (These considerations do not affect Church's primary argument—which we endorse—that the sentences 'Mary doubts that Vladimir believes attorneys are wealthy' and 'Mary doubts that Vladimir believes lawyers are wealthy' cannot differ in truth value in English if 'lawyer' and 'attorney' are synonyms.)

This interesting dispute between Church and Putnam is now over three decades old. The views they advocated then need not represent their current positions on these issues.

11. See also Ruth Barcan Marcus, 'A Proposed Solution to a Puzzle About Belief' in P. French, T. Uehling, and H. Wettstein (eds.), *Midwest Studies in Philosophy, vi: The Foundations of Analytic Philosophy* (University of Minnesota Press, 1981), 501–10, esp. at p. 505, and her 'Rationality and Believing the Impossible', *Journal of Philosophy* 80 (1983), 321–38, esp. p. 330. (Marcus argues here for the much stronger claim—which Richard rejects—that one cannot believe any necessary falsehood.) Other recent theories similarly committed to denying that King George has the contradictory belief that Scott is not Scott are given in David Lewis, 'What Puzzling Pierre Does not Believe', *Australasian Journal of Philosophy* 59 (1981), 283–9; and Robert Stalnaker, *Inquiry* (MIT Press/Bradford Books, 1984), at pp. 72–99, esp. at p. 85. The major criticism given in Soames's contributed article (ch. 11) applies to some extent to both of these theories.
12. One significant difference between Putnam and Richard is that Richard's analysis sanctions

(*i*) If *c* believes that . . *a* . . *a* . ., then *c* believes that . . *a* . . *b* . .

but not

(*ii*) If *c* believes that . . *a* . . *b* . ., then *c* believes that . . *a* . . *a* . .,

where *a* and *b* are expressions having the same content, while Putnam's fails to sanction either. In subsequent work both Richard (in 'Quantification and Leibniz's Law', *The Philosophical Review* 96 (1987), 555–78) and David Kaplan (in 'Word, Object, and Belief', unpublished lectures) have explored Putnam-like analyses that reject both (*i*) and (*ii*). These analyses are discussed in Soames, 'Substitutivity'.

I

KNOWLEDGE BY ACQUAINTANCE AND KNOWLEDGE BY DESCRIPTION

BERTRAND RUSSELL

THE object of the following paper is to consider what it is that we know in cases where we know propositions about 'the so-and-so' without knowing who or what the so-and-so is. For example, I know that the candidate who gets most votes will be elected, though I do not know who is the candidate who will get most votes. The problem I wish to consider is: What do we know in these cases, where the subject is merely described? I have considered this problem elsewhere[1] from a purely logical point of view; but in what follows I wish to consider the question in relation to theory of knowledge as well as in relation to logic, and in view of the above-mentioned logical discussions, I shall in this paper make the logical portion as brief as possible.

In order to make clear the antithesis between 'acquaintance' and 'description', I shall first of all try to explain what I mean by 'acquaintance'. I say that I am *acquainted* with an object when I have a direct cognitive relation to that object, that is when I am directly aware of the object itself. When I speak of a cognitive relation here, I do not mean the sort of relation which constitutes judgement, but the sort which constitutes presentation. In fact, I think the relation of subject and object which I call acquaintance is simply the converse of the relation of object and subject which constitutes presentation. That is, to say that *S* has acquaintance with *O* is essentially the same thing as to say that *O* is presented to *S*. But the associations and natural extensions of the word *acquaintance* are different from those of the word *presentation*. To

Bertrand Russell, 'Knowledge by Acquaintance and Knowledge by Description', *Proceedings of the Aristotelian Society* 11 (1910–11), 108–128.

begin with, as in most cognitive words, it is natural to say that I am acquainted with an object even at moments when it is not actually before my mind, provided it has been before my mind, and will be again whenever occasion arises. This is the same sense in which I am said to know that $2 + 2 = 4$ even when I am thinking of something else. In the second place, the word *acquaintance* is designed to emphasize, more than the word *presentation*, the relational character of the fact with which we are concerned. There is, to my mind, a danger that, in speaking of presentations, we may so emphasize the object as to lose sight of the subject. The result of this is either to lead to the view that there is no subject, whence we arrive at materialism; or to lead to the view that what is presented is part of the subject, whence we arrive at idealism, and should arrive at solipsism but for the most desperate contortions. Now I wish to preserve the dualism of subject and object in my terminology, because this dualism seems to me a fundamental fact concerning cognition. Hence I prefer the word *acquaintance*, because it emphasizes the need of a subject which is acquainted.

When we ask what are the kinds of objects with which we are acquainted, the first and most obvious example is *sense-data*. When I see a colour or hear a noise, I have direct acquaintance with the colour or the noise. The sense-datum with which I am acquainted in these cases is generally, if not always, complex. This is particularly obvious in the case of sight. I do not mean, of course, merely that the supposed physical object is complex, but that the direct sensible object is complex and contains parts with spatial relations. Whether it is possible to be aware of a complex without being aware of its constituents is not an easy question, but on the whole it would seem that there is no reason why it should not be possible. This question arises in an acute form in connection with self-consciousness, which we must now briefly consider.

In introspection, we seem to be immediately aware of varying complexes, consisting of objects in various cognitive and conative relations to ourselves. When I see the sun, it often happens that I am aware of my seeing the sun, in addition to being aware of the sun; and when I desire food, it often happens that I am aware of my desire for food. But it is hard to discover any state of mind in which I am aware of myself alone, as opposed to a complex of which I am a constituent. The question of the nature of self-consciousness is too large, and too slightly connected with our subject, to be argued at length here. It is, however, very difficult to account for plain facts if we assume that we do not have acquaintance with ourselves. It is plain that we are not only *acquainted* with the complex 'Self-acquainted-with-A', but we also *know* the proposition 'I am acquainted with A'. Now here the complex has been analysed, and if 'I' does not stand for something which is a direct object of acquaintance, we shall have to suppose that 'I' is something known by description. If we wished to maintain the view that there is no acquaintance with Self, we might argue as follows: We are acquainted with *acquaintance*, and we know that it is a relation. Also we are acquainted with a complex in which we perceive that acquaintance is the relating relation. Hence we know that this complex must have a constituent which is that which is acquainted, that is, must have a subject-term as well as an object-term. This subject-term we define as 'I'. Thus 'I' means 'the subject-term in awarenesses of which *I* am aware'. But as a definition this cannot be regarded as a happy effort. It would seem necessary, therefore, to suppose that I am acquainted with myself, and that 'I', therefore, requires no definition, being merely the proper name of a certain object. Thus self-consciousness cannot be regarded as throwing light on the question whether we can know a complex without knowing its constituents. This question, however, is not important for our present purposes, and I shall therefore not discuss it further.

The awarenesses we have considered so far have all been awarenesses of particular existents, and might all in a large sense be called sense-data. For, from the point of view of theory of knowledge, introspective knowledge is exactly on a level with knowledge derived from sight or hearing. But, in addition to awareness of the above kind of objects, which may be called awareness of *particulars*, we have also what may be called awareness of *universals*. Awareness of universals is called *conceiving*, and a universal of which we are aware is called a *concept*. Not only are we aware of particular yellows, but if we have seen a sufficient number of yellows and have sufficient intelligence, we are aware of the universal *yellow*; this universal is the subject in such judgements as 'yellow differs from blue' or 'yellow resembles blue less than green does'. And the universal yellow is the predicate in such judgements as 'this is yellow', where 'this' is a particular sense-datum. And universal relations, too, are objects of awarenesses; up and down, before and after, resemblance, desire, awareness itself, and so on, would seem to be all of them objects of which we can be aware.

In regard to relations, it might be urged that we are never aware of the universal relation itself, but only of complexes in which it is a constituent. For example, it may be said that we do not know directly such a relation as *before*, though we understand such a proposition as 'this is before that', and may be directly aware of such a complex as 'this being before that', This view, however, is difficult to reconcile with the fact that we often know propositions in which the relation is the subject, or in which the relata are not definite given objects but 'anything'. For example, we know that if one thing is before another, and the other before a third, then the first is before the third; and here the things concerned are not definite things, but 'anything'. It is hard to see how we could know such a fact about 'before' unless we were acquainted with 'before', and not merely with actual particular cases of one given object being before another given object. And more directly: A judgement such as 'this is before that', where this judgement is derived from awareness of a complex, constitutes an analysis, and we should not understand the analysis if we were not acquainted with the meaning of the terms employed. Thus we must suppose that we are acquainted with the meaning of 'before', and not merely with instances of it.

There are thus two sorts of objects of which we are aware, namely, particulars and universals. Among particulars I include all existents, and all complexes of which one or more constituents are existents, such as this-before-that, this-above-that, the-yellowness-of-this. Among universals I include all objects of which no particular is a constituent. Thus the disjunction 'universal–particular' is exhaustive. We might also call it the disjunction 'abstract–concrete'. It is not quite parallel with the opposition 'concept–percept', because things remembered or imagined belong with particulars, but can hardly be called percepts. (On the other hand, universals with which we are acquainted may be identified with concepts.)

It will be seen that among the objects with which we are acquainted are not included physical objects (as opposed to sense-data), nor other people's minds. These things are known to us by what I call 'knowledge by description', which we must now consider.

By a 'description' I mean any phrase of the form 'a so-and-so' or 'the so-and-so'. A phrase of the form 'a so-and-so' I shall call an 'ambiguous' description; a phrase of the form 'the so-and-so' (in the singular) I shall call a 'definite' description. Thus 'a man' is an ambiguous description, and 'the man with the iron mask' is a definite description. There are various problems connected with ambiguous descriptions, but I pass them by, since they do not directly concern the matter I wish to discuss. What I wish to discuss is the nature of our knowledge concerning objects in cases where we know that there is an object answering to a definite description, though we are not *acquainted* with any such object. This is a matter which is concerned exclusively with *definite* descriptions. I shall, therefore, in the sequel, speak simply of 'descriptions' when I mean 'definite descriptions'. Thus a description will mean any phrase of the form 'the so-and-so' in the singular.

I shall say that an object is 'known by description' when we know that it is '*the* so-and-so,' that is when we know that there is one object, and no more, having a certain property; and it will generally be implied that we do not have knowledge of the same object by acquaintance. We know that the man with the iron mask existed, and many propositions are known about him; but we do not know who he was. We know that the candidate who gets most votes will be elected, and in this case we are very likely also acquainted (in the only sense in which one can be acquainted with someone else) with the man who is, in fact, the candidate who will get most votes, but we do not know which of the candidates he is, that is we do not know any proposition of the form '*A* is the candidate who will get most votes' where *A* is one of the candidates by name. We shall say that we have '*merely* descriptive knowledge' of the so-and-so when, although we know that the so-and-so exists, and although we may possibly be acquainted with the object which is, in fact, the so-and-so, yet we do not know any proposition '*a* is the so-and-so', where *a* is something with which we are acquainted.

When we say 'the so-and-so exists', we mean that there is just one object which is the so-and-so. The proposition '*a* is the so-and-so' means that *a* has the property so-and-so, and nothing else has. 'Sir Joseph Larmor is the Unionist candidate' means 'Sir Joseph Larmor is a Unionist candidate, and no one else is'. 'The Unionist candidate exists' means 'someone is a Unionist candidate, and no one else is'. Thus, when we are acquainted with an object which is the so-and-so, we know that the so-and-so exists, but we may know that the so-and-so exists when we are not acquainted with any object which we know to be the so-and-so, and even when we are not acquainted with any object which, in fact, is the so-and-so.

Common words, even proper names, are usually really descriptions. That is to say, the thought in the mind of a person using a proper name correctly can generally only be expressed explicitly if we replace the proper name by a description. Moreover, the description required to express the thought will vary for different people, or for the same person at different times. The only thing constant (so long as the name is rightly used) is the object to which the name applies. But so long as this remains constant, the particular description involved usually makes no difference to the truth or falsehood of the proposition in which the name appears.

Let us take some illustrations. Suppose some statement made about Bismarck. Assuming that there is such a thing as direct acquaintance with oneself, Bismarck himself might have used his name directly to designate the particular person with whom he was acquainted. In this case, if he made a judgement about himself, he himself might be a constituent of the judgement. Here the proper name has the direct use which it always wishes to have, as simply standing for a certain object, and not for a description of the object. But if a person who knew Bismarck made a judgement about him, the case is different. What this person was acquainted with were certain sense-data which he connected (rightly, we will suppose) with Bismarck's body. His body as a physical object, and still more his mind, were only known as the body and the mind connected with these sense-data. That is, they were known by description. It is, of course, very much a matter of chance which characteristics of a man's appearance will come into a friend's mind when he thinks of him; thus the description actually in the friend's mind is accidental. The essential point is that he knows that the various descriptions all apply to the same entity, in spite of not being acquainted with the entity in question.

When we, who did not know Bismarck, make a judgement about him, the description in our minds will probably be some more or less vague mass of historical knowledge—far more, in most cases, than is required to identify him. But, for the sake of illustration, let us assume that we think of him as 'the first Chancellor of the German Empire'. Here all the words are abstract except 'German'. The word 'German' will again have different meanings for different people. To some it will recall travels in Germany, to some the look of Germany on the map, and so on. But if we are to obtain a description which we know to be applicable, we shall be compelled, at some point, to bring in a reference to a particular with which we are acquainted. Such reference is involved in any mention of past, present, and future (as opposed to definite dates), or of here and there, or of what others have told us. Thus it would seem that, in some way or other, a description known to be applicable to a particular must involve some reference to a particular with which we are acquainted, if our knowledge about the thing described is not to be merely what follows logically from the description. For example, 'the most long-lived of men' is a description which must apply to some man, but we can make no judgements concerning this man which involve knowledge about him beyond what the description gives. If, however, we say, 'the first Chancellor of the German Empire was an astute diplomatist', we can only be assured of the truth of our judgement in virtue of something with which we are acquainted—usually a testimony heard or read. Considered psychologically, apart from the information we convey to others, apart from the fact about the actual Bismarck, which gives importance to our judgement, the thought we really have contains the one or more particulars involved, and otherwise consists wholly of concepts. All names of places—London, England, Europe, the earth, the solar system—similarly involve, when used, descriptions which start from some one or more particulars with which we are acquainted. I suspect that even the universe, as considered by metaphysics, involves such a connection with particulars. In logic, on the contrary, where we are concerned not merely with what does exist, but with whatever might or could exist or be, no reference to actual particulars is involved.

It would seem that, when we make a statement about something only known by description, we often *intend* to make our statement, not in the form involving the description, but about the actual thing described. That is to say, when we say anything about Bismarck, we should like, if we could, to make the judgement which Bismarck alone can make, namely, the judgement of which he himself is a constituent. In this we are necessarily defeated, since the actual Bismarck is unknown to us. But we know that there is an object *B* called Bismarck, and that *B* was an astute diplomatist. We can thus *describe* the proposition we should like to affirm, namely '*B* was an astute diplomatist', where B is the object which was Bismarck. What enables us to communicate in spite of the varying descriptions we employ is that we know there is a true proposition concerning the actual Bismarck, and that however we may vary the proposition (so long as the description is correct), the proposition described is still the same. This proposition, which is described and is known to be true, is what interests us; but we are not acquainted with the proposition itself, and do not know *it*, though we know it is true.

It will be seen that there are various stages in the removal from acquaintance with particulars: there is Bismarck to people who knew him, Bismarck to those who only know of him through history, the man with the iron mask, the longest-lived of men. These are progressively further removed from acquaintance with particulars, and there is a similar hierarchy in the region of universals. Many universals, like many particulars, are only known to us by description. But here, as in the case of particulars, knowledge concerning what is known by description is ultimately reducible to knowledge concerning what is known by acquaintance.

The fundamental epistemological principle in the analysis of propositions containing descriptions is this: *Every proposition which we can understand must be composed wholly of constituents with which we are acquainted*. From what has been said already, it will be plain why I advocate this principle, and how I propose to meet the case of propositions which at first sight contravene it. Let us begin with the reasons for supposing the principle true.

The chief reason for supposing the principle true is that it seems scarcely possible to believe that we can make a judgement or entertain a supposition without knowing what it is that we are judging or supposing about. If we make a judgement about (say) Julius Caesar, it is plain that the actual person who was Julius Caesar is not a constituent of the judgement. But before going further, it may be well to explain what I mean when I say that this or that is a constituent of a judgement, or of a proposition which we understand. To begin with judgements: a judgement, as an occurrence, I take to be a relation of a mind to several entities, namely, the entities which compose what is judged. If, for example, I judge that *A* loves *B*, the judgement as an event consists in the existence, at a certain moment, of a specific four-term relation, called *judging*, between me and *A* and love and *B*. That is to say, at the time when I judge, there is a certain complex whose terms are myself and *A* and love and *B*, and whose relating relation is *judging*. (The relation *love* enters as one of the terms of the relation, not as a relating relation.) My reasons for this view have been set forth elsewhere,[2] and I shall not repeat them here. Assuming this view of judgement, the constituents of the judgement are simply the constituents of the complex which is the judgement. Thus, in the above case, the constituents are myself and *A* and love and *B* and judging. But myself and judging are constituents shared by all my judgements; thus the *distinctive* constituents of the particular judgement in question are *A* and love and *B*. Coming now to what is meant by 'understanding a proposition', I should say that there is another relation possible between me and *A* and love and *B*, which is called my *supposing* that *A* loves *B*.[3] When we can *suppose* that *A* loves *B*, we 'understand the proposition' *A loves B*. Thus we often understand a proposition in cases where we have not enough knowledge to make a judgement. Supposing, like judging, is a many-term relation, of which a mind is one term. The other terms of the relation are called the constituents of the proposition supposed. Thus the principle which I enunciated may be restated as follows: *Whenever a relation of supposing or judging occurs, the terms to which the supposing or judging mind is related by the relation of supposing or judging must be terms with which the mind in question is acquainted*. This is merely to say that we cannot make a judgement or a supposition without knowing what it is that we are making our judgement or supposition about. It seems to me that the truth of this principle is evident as soon as the principle is understood; I shall, therefore, in what follows, assume the principle, and use it as a guide in analysing judgements that contain descriptions.

Returning now to Julius Caesar, I assume that it will be admitted that he himself is not a constituent of any judgement which I can make. But at this point it is necessary to examine the view that judgements are composed of something called 'ideas', and that it is the 'idea' of Julius Caesar that is a constituent of my judgement. I believe the plausibility of this view rests upon a failure to form a right theory of descriptions. We may mean by my 'idea' of Julius Caesar the things that I know about him, for example that he conquered Gaul, was assassinated on the Ides of March, and is a plague to schoolboys. Now I am admitting, and indeed contending, that in order to discover what is actually in my mind when I judge about Julius Caesar, we must substitute for the proper name a description made up of some of the things I know about him. (A description which will often serve to express my thought is 'the man whose name was *Julius Caesar*'. For whatever else I may have forgotten about him, it is plain that when I mention him I have not forgotten that that was his name.) But although I think the theory that judgements consist of ideas may have been suggested in some such way, yet I think the theory itself is fundamentally mistaken. The view seems to be that there is some mental existent which may be called the 'idea' of something outside the mind of the person who has the idea, and that, since judgement is a mental event, its constituents must be constituents of the mind of the person judging. But in this view ideas become a veil between us and outside things—we never really, in knowledge, attain to the things we are supposed to be knowing about, but only to the ideas of those things. The relation of mind, idea, and object, on this view, is utterly obscure, and, so far as I can see, nothing discoverable by inspection warrants the intrusion of the idea between the mind and the object. I suspect that the view is fostered by the dislike of relations, and that it is felt the mind could not know objects unless there were something 'in' the mind which could be called the state of knowing the object. Such a view, however, leads at once to a vicious endless regress, since the relation of idea to object will have to be explained by supposing that the idea itself has an idea of the object, and so on *ad infinitum*. I therefore see no reason to believe that, when we are acquainted with an object, there is in us something which can be called the 'idea' of the object. On the contrary, I hold that acquaintance is wholly a relation, not demanding any such constituent of the mind as is supposed by advocates of 'ideas'. This is, of course, a large question, and one which would take us far from our subject if it were adequately discussed. I therefore content myself with the above indications, and with the corollary that, in judging, the actual objects concerning which we judge, rather than any supposed purely mental entities, are constituents of the complex which is the judgement.

When, therefore, I say that we must substitute for 'Julius Caesar' some description of Julius Caesar, in order to discover the meaning of a judgement nominally about him, I am not saying that we must substitute an idea. Suppose our description is 'the man whose name was *Julius Caesar*'. Let our judgement be 'Julius Caesar was assassinated'. Then it becomes 'the man whose name was *Julius Caesar* was assassinated'. Here *Julius Caesar* is a noise or shape with which we are acquainted, and all the other constituents of the judgement (neglecting the tense in 'was') are *concepts* with which we are acquainted. Thus our judgement is wholly reduced to constituents with which we are acquainted, but Julius Caesar himself has ceased to be a constituent of our judgement. This, however, requires a proviso, to be further explained shortly, namely, that 'the man whose name was *Julius Caesar*' must not, as a whole, be a constituent of our judgement, that is to say, this phrase must not, as a whole, have a meaning which enters into the judgement. Any right analysis of the judgement, therefore, must break up this phrase, and not treat it as a subordinate complex which is part of the judgement. The judgement 'the man whose name was *Julius Caesar* was assassinated' may be interpreted as meaning 'One and only one man was called *Julius Caesar*, and that one was assassinated'. Here it is plain that there is no constituent corresponding to the phrase 'the man whose name was *Julius Caesar*'. Thus there is no reason to regard this phrase as expressing a constituent of the judgement, and we have seen that this phrase must be broken up if we are to be acquainted with all the constituents of the judgement. This conclusion, which we have reached from considerations concerned with the theory of knowledge, is also forced upon us by logical considerations, which must now be briefly reviewed.

It is common to distinguish two aspects, *meaning* and *denotation*, in such phrases as 'the author of *Waverley*'. The meaning will be a certain complex, consisting (at least) of authorship and *Waverley* with some relation; the denotation will be Scott. Similarly, 'featherless bipeds' will have a complex meaning, containing as constituents the presence of two feet and the absence of feathers, while its denotation will be the class of men. Thus when we say 'Scott is the author of *Waverley*' or 'men are the same as featherless bipeds', we are asserting an identity of denotation, and this assertion is worth making because of the diversity of meaning.[4] I believe that the duality of meaning and denotation, though capable of a true interpretation, is misleading if taken as fundamental. The denotation, I believe, is not a constituent of the proposition, except in the case of proper names, that is of words which do not assign a property to an object, but merely and solely name it. And I should hold further that, in this sense, there are only two words which are strictly proper names of particulars, namely, 'I' and 'this.'

One reason for not believing the denotation to be a constituent of the proposition is that we may know the proposition even when we are not acquainted with the denotation. The proposition 'the author of *Waverley* is a novelist' was known to people who did not know that 'the author of *Waverley*' denoted Scott. This reason has been already sufficiently emphasized.

A second reason is that propositions concerning 'the so-and-so' are possible even when 'the so-and-so' has no denotation. Take, for example 'the golden mountain does not exist' or 'the round square is self-contradictory'. If we are to preserve the duality of meaning and denotation, we have to say, with Meinong, that there are such objects as the golden mountain and the round square, although these objects do not have being. We even have to admit that the existent round square is existent, but does not exist.[5] Meinong does not regard this as a contradiction, but I fail to see that it is not one. Indeed, it seems to me evident that the judgement 'there is no such object as the round square' does not presuppose that there is such an object. If this is admitted, however, we are led to the conclusion that, by parity of form, no judgement concerning 'the so-and-so' actually involves the so-and-so as a constituent.

Miss Jones[6] contends that there is no difficulty in admitting contradictory predicates concerning such an object as 'the present King of France', on the ground that this object is in itself contradictory. Now it might, of course, be argued that this object, unlike the round square, is not self-contradictory, but merely non-existent. This, however, would not go to the root of the matter. The real objection to such an argument is that the law of contradiction ought not to be stated in the traditional form '*A* is not both *B* and not *B*', but in the form 'no proposition is both true and false'. The traditional form applies only to certain propositions, namely, to those which attribute a predicate to a subject. When the law is stated of propositions, instead of being stated concerning subjects and predicates, it is at once evident that propositions about the present King of France or the round square can form no exception, but are just as incapable of being both true and false as other propositions.

Miss Jones[7] argues that 'Scott is the author of *Waverley*' asserts identity of denotation between *Scott* and *the author of Waverley*. But there is some difficulty in choosing among alternative meanings of this contention. In the first place, it should be observed that *the author of Waverley* is not a *mere* name, like *Scott*. *Scott* is merely a noise or shape conventionally used to designate a certain person; it gives us no information about that person, and has nothing that can be called meaning as opposed to denotation. (I neglect the fact, considered above, that even proper names, as a rule, really stand for descriptions.) But *the author of Waverley* is not merely conventionally a name for Scott; the element of mere convention belongs here to the separate words, *the* and *author* and *of* and *Waverley*. Given what these words stand for, *the author of Waverley* is no longer arbitrary. When it is said that Scott is the author of *Waverley*, we are not stating that these are two *names* for one man, as we should be if we said 'Scott is Sir Walter'. A man's name is what he is called, but however much Scott had been called the author of *Waverley*, that would not have made him be the author; it was necessary for him actually to write *Waverley*, which was a fact having nothing to do with names.

If, then, we are asserting identity of denotation, we must not mean by *denotation* the mere relation of a name to the thing named. In fact, it would be nearer to the truth to say that the *meaning* of 'Scott' is the *denotation* of 'the author of *Waverley*'. The relation of 'Scott' to Scott is that 'Scott' means Scott, just as the relation of 'author' to the concept which is so called is that 'author' means this concept. Thus if we distinguish meaning and denotation in 'the author of *Waverley*', we shall have to say that 'Scott' has meaning but not denotation. Also when we say 'Scott is the author of *Waverley*', the *meaning* of 'the author of *Waverley*' is relevant to our assertion. For if the denotation alone were relevant, any other phrase with the same denotation would give the same proposition. Thus 'Scott is the author of *Marmion*' would be the same proposition as 'Scott is the author of *Waverley*'. But this is plainly not the case, since from the first we learn that Scott wrote *Marmion* and from the second we learn that he wrote *Waverley*, but the first tells us nothing about *Waverley* and the second nothing about *Marmion*. Hence the meaning of 'the author of *Waverley*', as oppsed to the denotation, is certainly relevant to 'Scott is the author of *Waverley*'.

We have thus agreed that 'the author of *Waverley*' is not a mere name, and that its meaning is relevant in propositions in which it occurs. Thus if we are to say, as Miss Jones does, that 'Scott is the author of *Waverley*' asserts an identity of denotation, we must regard the denotation of 'the author of *Waverley*' as the denotation of what is *meant* by 'the author of *Waverley*'. Let us call the meaning of 'the author of *Waverley*' *M*. Thus *M* is what 'the author of *Waverley*' means. Then we are to suppose that 'Scott is the author of *Waverley*' means 'Scott is the denotation of *M*'. But here we are explaining our proposition by another of the same form, and thus we have made no progress towards a real explanation. 'The denotation of *M*', like 'the author of *Waverley*', has both meaning and denotation, on the theory we are examining. If we call its meaning *M'*, our proposition becomes 'Scott is the denotation of *M'*'. But this leads at once to an endless regress. Thus the attempt to regard our proposition as asserting identity of denotation breaks down, and it becomes imperative to find some other analysis. When this analysis has been completed, we shall be able to reinterpret the phrase 'identity of denotation', which remains obscure so long as it is taken as fundamental.

The first point to observe is that, in any proposition about 'the author of *Waverley*', provided Scott is not explicitly mentioned, the denotation itself, that is Scott, does not occur, but only the concept of denotation, which will be represented by a variable. Suppose we say 'the author of *Waverley* was the author of *Marmion*', we are certainly not saying that both were Scott—we may have forgotten that there was such a person as Scott. We are saying that there is some man who was the author of *Waverley* and the author of *Marmion*. That is to say, there is someone who wrote *Waverley* and *Marmion*, and no one else wrote them. Thus the identity is that of a variable, that is of an indefinite subject, 'someone'. This is why we can understand propositions about 'the author of *Waverley*', without knowing who he was. When we say 'the author of *Waverley* was a poet' we mean 'one and only one man wrote *Waverley*, and he was a poet'; when we say 'the author of *Waverley* was Scott' we mean 'one and only one man wrote *Waverley* and he was Scott'. Here the identity is between a variable, that is an indeterminate subject ('he'), and Scott; 'the author of *Waverley*' has been analysed away, and no longer appears as a constituent of the proposition.[8]

The reason why it is imperative to analyse away the phrase 'the author of *Waverley*' may be stated as follows. It is plain that when we say 'the author of *Waverley* is the author of *Marmion*', the *is* expresses identity. We have seen also that the common *denotation*, namely Scott, is not a constituent of this proposition, while the *meanings* (if any) of 'the author of *Waverley*' and 'the author of

Marmion' are not identical. We have seen also that, in any sense in which the meaning of a word is a constituent of a proposition in whose verbal expression the word occurs, 'Scott' means the actual man Scott, in the same sense in which 'author' means a certain universal. Thus, if 'the author of *Waverley*' were a subordinate complex in the above proposition, its *meaning* would have to be what was said to be identical with the *meaning* of 'the author of *Marmion*'. This is plainly not the case; and the only escape is to say that 'the author of *Waverley*' does not, by itself, have a meaning, though phrases of which it is part do have a meaning. That is, in a right analysis of the above proposition, 'the author of *Waverley*' must disappear. This is effected when the above proposition is analysed as meaning: 'Someone wrote *Waverley* and no one else did, and that someone also wrote *Marmion* and no one else did'. This may be more simply expressed by saying that the propositional function 'x wrote *Waverley* and *Marmion*, and no one else did' is capable of truth, that is some value of x makes it true. Thus the true subject of our judgement is a propositional function, that is a complex containing an undetermined constituent, and becoming a proposition as soon as this constituent is determined.

We may now define the denotation of a phrase. If we know that the proposition 'a is the so-and-so' is true, that is that a is so-and-so and nothing else is, we call a the denotation of the phrase 'the so-and-so'. A very great many of the propositions we naturally make about 'the so-and-so' will remain true or remain false if we substitute a for 'the so-and-so', where a is the denotation of 'the so-and-so'. Such propositions will also remain true or remain false if we substitute for 'the so-and-so' any other phrase having the same denotation. Hence, as practical men, we become interested in the denotation more than in the description, since the denotation decides as to the truth or falsehood of so many statements in which the description occurs. Moreover, as we saw earlier in considering the relations of description and acquaintance, we often wish to reach the denotation, and are only hindered by lack of acquaintance: in such cases the description is merely the means we employ to get as near as possible to the denotation. Hence it naturally comes to be supposed that the denotation is part of the proposition in which the description occurs. But we have seen, both on logical and on epistemological grounds, that this is an error. The actual object (if any) which is the denotation is not (unless it is explicitly mentioned) a constituent of proposition in which descriptions occur; and this is the reason why, in order to understand such propositions, we need acquaintance with the constituents of the description, but do not need acquaintance with its denotation. The first result of analysis, when applied to propositions whose grammatical subject is 'the so-and-so', is to substitute a variable as subject; that is we obtain a proposition of the form 'There is *something* which alone is so-and-so, and that *something* is such-and-such'. The further analysis of propositions concerning 'the so-and-so' is thus merged in the problem of the nature of the variable, that is of the meanings of *some*, *any*, and *all*. This is a difficult problem, concerning which I do not intend to say anything at present.

To sum up our whole discussion: We began by distinguishing two sorts of knowledge of objects, namely, knowledge by *acquaintance* and knowledge by *description*. Of these it is only the former that brings the object itself before the mind. We have acquaintance with sense-data, with many universals, and possibly with ourselves, but not with physical objects or other minds. We have *descriptive* knowledge of an object when we know that it is *the* object having some property or properties with which we are acquainted; that is to say, when we know that the property or properties in question belong to one object and no more, we are said to have knowledge of that one object by description, whether or not we are acquainted with the object. Our knowledge of physical objects and of other minds is only knowledge by description, the descriptions involved being usually such as involve sense-data. All propositions intelligible to us, whether or not they primarily concern things only known to us by description, are composed wholly of constituents with which we are acquainted, for a constituent with which we are not acquainted is unintelligible to us. A judgement, we found, is not composed of mental constituents called 'ideas', but consists of a complex whose constituents are a mind and certain objects, particulars or universals. (One at least must be a universal.) When a judgement is rightly analysed, the objects which are constituents of it must all be objects with which the mind which is a constituent of it is acquainted. This conclusion forces us to analyse descriptive phrases occurring in propositions, and to say that the objects denoted by such phrases are not constituents of judgements in which such phrases occur (unless these objects are explicitly mentioned). This leads us to the view (recommended also on purely logical grounds) that when we say 'the author of *Marmion* was the author of *Waverley*', Scott himself is not a constituent of our judgement, and that the judgement cannot be explained by saying that it affirms identity of denotation with diversity of connotation. It also, plainly, does not assert identity of meaning. Such judgements, therefore, can only be analysed by breaking up the descriptive phrases, introducing a variable, and making propositional functions the ultimate subjects. In fact, 'the so-and-so is such-and-such' will mean that '*x* is so-and-so and nothing else is, and *x* is such-and-such' is capable of truth. The analysis of such judgements involves many fresh problems, but the discussion of these problems is not undertaken in the present paper.

Notes

1. See references below.
2. 'The Nature of Truth' in *Philosophical Essays*.
3. Cf. Meinong, *Ueber Annahmen, passim*. I formerly supposed, contrary to Meinong's view, that the relationship of supposing might be merely that of presentation. In this view I now think I was mistaken and Meinong is right. But my present view depends upon the theory that both in judgement and in assumption there is no single Objective, but the several constituents of the judgement or assumption are in a many-term relation to the mind.
4. This view has been recently advocated by Miss E. E. C. Jones, 'A New Law of Thought and its Implications', *Mind* (January, 1911).
5. Meinong, *Ueber Annahmen*, 2nd ed. (Leipzig, 1910), 141.
6. *Mind* (July, 1910), p. 380.
7. *Mind* (July,1910), p. 379.
8. The theory which I am advocating is set forth fully, with the logical grounds in its favour, in *Principia Mathematica*, vol. I, Introduction, chap. III; also, less fully, in *Mind* (October, 1905).

II

THOUGHTS

GOTTLOB FREGE

JUST as 'beautiful' points the ways for aesthetics and 'good' for ethics, so do words like 'true' for logic. All sciences have truth as their goal; but logic is also concerned with it in a quite different way: logic has much the same relation to truth as physics has to weight or heat. To discover truths is the task of all sciences; it falls to logic to discern the laws of truth. The word 'law' is used in two senses. When we speak of moral or civil laws we mean prescriptions which ought to be obeyed but with which actual occurrences are not always in conformity. Laws of nature are general features of what happens in nature, and occurrences in nature are always in accordance with them. It is rather in this sense that I speak of laws of truth. Here of course it is not a matter of what happens but of what is. From the laws of truth there follow prescriptions about asserting, thinking, judging, inferring. And we may very well speak of laws of thought in this way too. But there is at once a danger here of confusing different things. People may very well interpret the expression 'law of thought' by analogy with 'law of nature' and then have in mind general features of thinking as a mental occurrence. A law of thought in this sense would be a psychological law. And so they might come to believe that logic deals with the mental process of thinking and with the psychological laws in accordance with which this takes place. That would be misunderstanding the task of logic, for truth has not here been given its proper place. Error and superstition have causes just as much as correct cognition. Whether what you take for true is false

Gottlob Frege, 'Thoughts', in G. Frege, *Collected Papers on Mathematics, Logic, and Philosophy* (Oxford: Basil Blackwell, 1984), ed. B. McGuinness, trans. P. Geach and R. H. Stoothoff, 351–72.

(Alternative translation entitled 'The Thought: A Logical Inquiry' in P. F. Strawson (ed.), *Philosophical Logic*, (Oxford Readings in Philosophy, 1976), 17–38.)

or true, your so taking it comes about in accordance with psychological laws. A derivation from these laws, an explanation of a mental process that ends in taking something to be true, can never take the place of proving what is taken to be true. But may not logical laws also have played a part in this mental process? I do not want to dispute this, but if it is a question of truth this possibility is not enough. For it is also possible that something non-logical played a part in the process and made it swerve from the truth. We can decide only after we have come to know the laws of truth; but then we can probably do without the derivation and explanation of the mental process, if our concern is to decide whether the process terminates in *justifiably* taking something to be true. In order to avoid any misunderstanding and prevent the blurring of the boundary between psychology and logic, I assign to logic the task of discovering the laws of truth, not the laws of taking things to be true or of thinking. The meaning of the word 'true' is spelt out in the laws of truth.

But first I shall attempt to outline roughly how I want to use 'true' in this connection, so as to exclude irrelevant uses of the word. 'True' is not to be used here in the sense of 'genuine' or 'veracious'; nor yet in the way it sometimes occurs in discussion of artistic questions, when, for example, people speak of truth in art, when truth is set up as the aim of art, when the truth of a work of art or true feeling is spoken of. Again, the word 'true' is prefixed to another word in order to show that the word is to be understood in its proper, unadulterated sense. This use too lies off the path followed here. What I have in mind is that sort of truth which it is the aim of science to discern.

Grammatically, the word 'true' looks like a word for a property. So we want to delimit more closely the region within which truth can be predicated, the region in which there is any question of truth. We find truth predicated of pictures, ideas, sentences, and thoughts. It is striking that visible and audible things turn up here along with things which cannot be perceived with the senses. This suggests that shifts of meaning have taken place. So indeed they have! Is a picture considered as a mere visible and tangible thing really true, and a stone or a leaf not true? Obviously we could not call a picture true unless there were an intention involved. A picture is meant to represent something. (Even an idea is not called true in itself, but only with respect to an intention that the idea should correspond to something.) It might be supposed from this that truth consists in a correspondence of a picture to what it depicts. Now a correspondence is a relation. But this goes against the use of the word 'true', which is not a relative term and contains no indication of anything else to which something is to correspond. If I do not know that a picture is meant to represent Cologne Cathedral then I do not know what to compare the picture with in order to decide on its truth. A correspondence, moreover, can only be perfect if the corresponding things coincide and so just are not different things. It is supposed to be possible to test the genuineness of a banknote by comparing it stereoscopically with a genuine one. But it would be ridiculous to try to compare a gold piece stereoscopically with a twenty-mark note. It would only be possible to compare an idea with a thing if the thing were an idea too. And then, if the first did correspond perfectly with the second, they would coincide. But this is not at all what people intend when they define truth as the correspondence of an idea with something real. For in this case it is essential precisely that the reality shall be distinct from the idea. But then there can be no complete correspondence, no complete truth. So nothing at all would be true: for what is only half-true is untrue. Truth does not admit of more or less. But could we not maintain that there is truth when there is correspondence in a certain respect? But which respect? For in that case what ought we to do so as to decide whether something is true? We should have to inquire whether it is *true* that an idea and a reality, say, correspond in the specified respect. And then we should be confronted by a question of the same kind, and the game could begin again. So the attempted explanation of truth as correspondence breaks down. And any other attempt to define truth also breaks down. For in a definition certain characteristics would have to be specified. And in application to any particular case the question would always arise whether it were *true* that the characteristics were present. So we should be going round in a circle. It therefore seems likely that the content of the word 'true' is *sui generis* and indefinable.

When we ascribe truth to a picture we do not really mean to ascribe a property which would belong to this picture quite independently of other things; we always have in mind some totally different object and we want to say that the picture corresponds in some way to this object. 'My idea corresponds to Cologne Cathedral' is a sentence, and now it is a matter of the truth of this sentence. So what is improperly called the truth of pictures and ideas is reduced to the truth of sentences. What is it that we call a sentence? A series of sounds, but only if it has a sense (this is not meant to convey that *any* series of sounds that has a sense is a sentence). And when we call a sentence true we really mean that its sense is true. And hence the only thing that raises the question of truth at all is the sense of sentences. Now is the sense of a sentence an idea? In any case, truth does not consist in correspondence of the sense with something else, for otherwise the question of truth would be reiterated to infinity.

Without offering this as a definition, I mean by 'a thought' something for which the question of truth can arise at all. So I count what is false among thoughts no less than what is true.[1] I can therefore say: thoughts are senses of sentences, without wishing to assert that the sense of every sentence is a thought. The thought, in itself imperceptible by the senses, becomes clothed in the perceptible garb of a sentence, and thereby we are enabled to grasp it. We say a sentence *expresses* a thought.

A thought is something imperceptible: anything the senses can perceive is excluded from the realm of things for which the question of truth arises. Truth is not a quality that answers to a particular kind of sense-impression. So it is sharply distinguished from the qualities we call by the names 'red', 'bitter', 'lilac-smelling'. But do we not see that the Sun has risen? And do we not then also see that this is true? That the Sun has risen is not an object emitting rays that reach my eyes; it is not a visible thing like the Sun itself. That the Sun has risen is recognized to be true on the basis of sense-impressions. But being true is not a sensible, perceptible property. A thing's being magnetic is also recognized on the basis of sense-impressions of the thing, although this property does not answer, any more than truth does, to a particular kind of sense-impression. So far these properties agree. However, we do need sense-impressions in order to recognize a body as magnetic. On the other hand, when I find it is true that I do not smell anything at this moment, I do not do so on the basis of sense-impressions.

All the same it is something worth thinking about that we cannot recognize a property of a thing without at the same time finding the thought *this thing has this property* to be true. So with every property of a thing there is tied up a property of a thought, namely truth. It is also worth noticing that the sentence 'I smell the scent of violets' has just the same content as the sentence 'It is true that I smell the scent of violets.' So it seems, then, that nothing is added to the thought by my ascribing to it the property of truth. And yet is it not a great result when the scientist after much hesitation and laborious research can finally say 'My conjecture is true'? The meaning of the word 'true' seems to be altogether *sui generis*. May we not be dealing here with something which cannot be called a property in the ordinary sense at all? In spite of this doubt I shall begin by expressing myself in accordance with ordinary usage, as if truth were a property, until some more appropriate way of speaking is found.

In order to bring out more precisely what I mean by 'a thought', I shall distinguish various kinds of sentences.[2] We should not wish to deny sense to a command, but this sense is not such that the question of truth could arise for it. Therefore I shall not call the sense of a command a thought. Sentences expressing wishes or requests are ruled out in the same way. Only those sentences in which we communicate or assert something come into the question. But here I do not count exclamations in which one vents one's feelings, groans, sighs, laughs—unless it has been decided by some special convention that they are to communicate something. But what about interrogative sentences? In a word-question[3] we utter an incomplete sentence, which is meant to be given a true sense just by means of the completion for which we are asking. Word-questions are accordingly left out of consideration here. Propositional questions[4] are a different matter. We expect to hear 'yes' or 'no'. The answer 'yes' means the same as an assertoric sentence, for in saying 'yes' the speaker presents as true the thought that was already completely contained in the interrogative sentence. This is how a propositional question can be formed from any assertoric sentence. And this is why an exclamation cannot be regarded as a communication: no corresponding propositional question can be formed. An interrogative sentence and an assertoric one contain the same thought; but the assertoric sentence contains something else as well, namely assertion. The interrogative sentence contains something more too, namely a request. Therefore two things must be distinguished in an assertoric sentence: the content, which it has in common with the corresponding propositional question; and assertion. The former is the thought or at least contains the thought. So it is possible to express a thought without laying it down as true. The two things are so closely joined in an assertoric sentence that it is easy to overlook their separability. Consequently we distinguish:

(1) the grasp of a thought—thinking,
(2) the acknowledgement of the truth of a thought—the act of judgement,[5]
(3) the manifestation of this judgement—assertion.

We have already performed the first act when we form a propositional question. An advance in science usually takes place in this way: first a thought is grasped, and thus may perhaps be expressed in a propositional question; after appropriate investigations, this thought is finally recognized to be true. We express acknowledgement of truth in the form of an assertoric sentence. We do not need the word 'true' for this. And even when we do use it the properly assertoric force does not lie in it, but in the assertoric sentence-form; and where this form loses its assertoric force the word 'true' cannot put it back again. This happens when we are not speaking seriously. As stage thunder is only sham thunder and a stage fight only a sham fight, so stage assertion is only sham assertion. It is only acting, only fiction. When playing his part the actor is not asserting anything; nor is he lying, even if he says something of whose falsehood he is convinced. In poetry we have the case of thoughts being expressed without being actually put forward as true, in spite of the assertoric form of the sentence; although the poem may suggest to the hearer that he himself should make an assenting judgement. Therefore the question still arises, even about what is presented in the assertoric sentence-form, whether it really contains an assertion. And this question must be answered in the negative if the requisite seriousness is lacking. It is unimportant whether the word 'true' is used here. This explains why it is that nothing seems to be added to a thought by attributing to it the property of truth.

An assertoric sentence commonly contains, over and above a thought and assertion, a third component not covered by the assertion. This is often meant to act on the feelings and mood of the hearer, or to arouse his imagination. Words like 'regrettably' and 'fortunately' belong here. Such constituents of sentences are more strongly prominent in poetry, but are seldom wholly absent from prose. They occur more rarely in mathematical, physical, or chemical expositions than in historical ones. What are called the humanities are closer to poetry, and are therefore less scientific, than the exact sciences, which are drier in proportion to being more exact; for exact science is directed toward truth and truth alone. Therefore all constituents of sentences not covered by the assertoric force do not belong to scientific exposition; but they are sometimes hard to avoid, even for one who sees the danger connected with them. Where the main thing is to approach by way of intimation what cannot be conceptually grasped, these constituents are fully justified. The more rigorously scientific an exposition is, the less the nationality of its author will be discernible and the easier it will be to translate. On the other hand, the constituents of language to which I here want to call attention make the translation of poetry very difficult, indeed make perfect translation almost always impossible, for it is just in what largely makes the poetic value that languages most differ.

It makes no difference to the thought whether I use the word 'horse' or 'steed' or 'nag' or 'prad'. The assertoric force does not cover the ways in which these words differ. What is called mood, atmosphere, illumination in a poem, what is portrayed by intonation and rhythm, does not belong to the thought.

Much in language serves to aid the hearer's understanding, for instance emphasizing part of a sentence by stress or word order. Here let us bear in mind words like 'still' and 'already'. Someone using the sentence 'Alfred has still not come' actually says 'Alfred has not come', and at the same time hints—but only hints—that Alfred's arrival is expected. Nobody can say: Since Alfred's arrival is not expected, the sense of the sentence is false. The way that 'but' differs from 'and' is that we use it to intimate that what follows it contrasts with what was to be expected from what preceded it. Such conversational suggestions make no difference to the thought. A sentence can be transformed by changing the verb from active to passive and at the same time making the accusative into the subject. In the same way we may change the dative into the nominative and at the same time replace 'give' by 'receive'. Naturally such transformations are not trivial in every respect; but they do not touch the thought, they do not touch what is true or false. If the inadmissibility of such transformations were recognized as a principle, then any profound logical investigation would be hindered. It is just as important to ignore distinctions that do not touch the heart of the matter, as to make distinctions which concern essentials. But what is essential depends on one's purpose. To a mind concerned with the beauties of language, what is trivial to the logician may seem to be just what is important.

Thus the content of a sentence often goes beyond the thought expressed by it. But the opposite often happens too; the mere wording which can be made permanent by writing or the gramophone, does not suffice for the expression of the thought. The present tense is used in two ways: first, in order to indicate a time; second, in order to eliminate any temporal restriction, where timelessness or eternity is part of the thought—consider for instance the laws of mathematics. Which of the two cases occurs is not expressed but must be devined. If a time-indication is conveyed by the present tense one must know when the sentence was uttered in order to grasp the thought correctly. Therefore the time of utterance is part of the expression of the thought. If someone wants to say today what he expressed yesterday using the word 'today', he will replace this word with 'yesterday'. Although the thought is the same, its verbal expression must be different in order that the change of sense which would otherwise be effected by the differing times of utterance may be cancelled out. It is the same with words like 'here' and 'there'. In all such cases the mere wording, as it can be preserved in writing, is not the complete expression of the thought; the knowledge of certain conditions accompanying the utterance, which are used as means of expressing the thought, is needed for us to grasp the thought correctly. Pointing the finger, hand gestures, glances may belong here too. The same utterance containing the word 'I' in the mouths of different men will express different thoughts, of which some may be true, others false.

The occurrence of the word 'I' in a sentence gives rise to some further questions.

Consider the following case. Dr Gustav Lauben says, 'I was wounded'. Leo Peter hears this and remarks some days later, 'Dr Gustav Lauben was wounded.' Does this sentence express the same thought as the one Dr Lauben uttered himself? Suppose that Rudolph Lingens was present when Dr Lauben spoke and now hears what is related by Leo Peter. If the same thought was uttered by Dr Lauben and Leo Peter, then Rudolph Lingens, who is fully master of the language and remembers what Dr Lauben said in his presence, must now know at once from Leo Peter's report that he is speaking of the same thing. But knowledge of the language is a special thing when proper names are involved. It may well be the case that only a few people associate a definite thought with the sentence 'Dr Lauben was wounded.' For complete understanding one needs in this case to know the expression 'Dr Gustav Lauben'. Now if both Leo Peter and Rudolph Lingens mean by 'Dr Gustav Lauben', the doctor who is the only doctor living in a house known to both of them, then they both understand the sentence 'Dr Gustav Lauben was wounded' in the same way; they associate the same thought with it. But it is also possible that Rudolph Lingens does not know Dr Lauben personally and does not know that it was Dr Lauben who recently said 'I was wounded'. In this case Rudolph Lingens cannot know that the same affair is in question. I say, therefore, in this case: the thought which Leo Peter expresses is not the same as that which Dr Lauben uttered.

Suppose further that Herbert Garner knows that Dr Gustav Lauben was born on 13 September 1875 in N.N. and this is not true of anyone else; suppose, however, that he does not know where Dr Lauben now lives nor indeed anything else about him. On the other hand, suppose Leo Peter does not know that Dr Lauben was born on 13 September 1875, in N.N. Then as far as the proper name 'Dr Gustav Lauben' is concerned, Herbert Garner and Leo Peter do not speak the same language, although they do in fact refer to the same man with this name; for they do not know that they are doing so. Therefore Herbert Garner does not associate the same thought with the sentence 'Dr Gustav Lauben was wounded' as Leo Peter wants to express with it. To avoid the awkwardness that Herbert Garner and Leo Peter are not speaking the same language, I shall suppose that Leo Peter uses the proper name 'Dr Lauben' and Herbert Garner uses the proper name 'Gustav Lauben'. Then it is possible that Herbert Garner takes the sense of the sentence 'Dr Lauben was wounded' to be true but is misled by false information into taking the sense of the sentence 'Gustav Lauben was wounded' to be false. So given our assumptions these thoughts are different.

Accordingly, with a proper name, it is a matter of the way that the object so designated is presented. This may happen in different ways, and to every such way there corresponds a special sense of a sentence containing the proper name. The different thoughts thus obtained from the same sentences correspond in truth-value, of course; that is to say, if one is true than all are true, and if one is false then all are false. Nevertheless, the difference must be recognized. So we must really stipulate that for every proper name there shall be just one associated manner of presentation of the object so designated. It is often unimportant that this stipulation should be fulfilled, but not always.

Now every one is presented to himself in a special and primitive way in which he is presented to no one else. So, when Dr Lauben has the thought that he was wounded, he will probably be basing it on this primitive way in which he is presented to himself. And only Dr Lauben himself can grasp thoughts specified in this way. But now he may want to communicate with others. He cannot communicate a thought he alone can grasp. Therefore, if he now says 'I was wounded', he must use 'I' in a sense which can be grasped by others, perhaps in the sense of 'he who is speaking to you at this moment'; by doing this he makes the conditions accompanying his utterance serve towards the expression of a thought.[6]

Yet there is a doubt: Is it at all the same thought which first that man expresses and then this one?

A man who is still unaffected by philosophy first of all gets to know things he can see and touch, can in short perceive with the senses, such as trees, stones, and houses, and he is convinced that someone else can equally see and touch the same tree and the same stone as he himself sees and touches. Obviously a thought does not belong with these things. Now can it, nevertheless, like a tree be presented to people as identical?

Even an unphilosophical man soon finds it necessary to recognize an inner world distinct from the outer world, a world of sense-impressions, of creations of his imagination, of sensations, of feelings and moods, a world of inclinations, wishes, and decisions. For brevity's sake I want to use the word 'idea' to cover all these occurrences, except decisions.

Now do thoughts belong to this inner world? Are they ideas? They are obviously not decisions.

How are ideas distinct from the things of the outer world?

First: ideas cannot be seen, or touched, or smelt, or tasted, or heard.

I go for a walk with a companion. I see a green field, I thus have a visual impression of the green. I have it, but I do not see it.

Secondly: ideas are something we have. We have sensations, feelings, moods, inclinations, wishes. An idea that someone has belongs to the content of his consciousness.

The field and the frogs in it, the Sun which shines on them, are there no matter whether I look at them or not, but the sense-impression I have of green exists only because of me — I am its owner. It seems absurd to us that a pain, a mood, a wish should go around the world without an owner, independently. A sensation is impossible without a sentient being. The inner world presupposes somebody whose inner world it is.

Thirdly: ideas need an owner. Things of the outer world are on the contrary independent.

My companion and I are convinced that we both see the same field; but each of us has a particular sense-impression of green. I glimpse a strawberry among the green strawberry leaves. My companion cannot find it, he is colour-blind. The colour-impression he gets from the strawberry is not noticeably different from the one he gets from the leaf. Now does my companion see the green leaf as red, or does he see the red berry as green, or does he see both with one colour which I am not acquainted with at all? These are unanswerable, indeed really nonsensical, questions; because when the word 'red' is meant not to state a property of things but to characterize sense-impressions belonging to my consciousness, it is only applicable within the realm of my consciousness. For it is impossible to compare my sense-impression with someone else's. In order to do that it would be necessary to bring together in one consciousness a sense-impression belonging to one consciousness and a sense-impression belonging to another consciousness. Now even if it were possible to make an idea disappear from one consciousness and at the same time make an idea appear in another consciousness, the question whether it is the same idea would still remain unanswerable. It is so much of the essence of any one of my ideas to be a content of my consciousness, that any idea someone else has is, just as such, different from mine. But might it not be possible that my ideas, the entire content of my consciousness, might be at the same time the content of a more embracing, perhaps Divine consciousness? Only if I were myself part of the Divine Being. But then would they really be my ideas, would I be their owner? This so far oversteps the limits of human understanding that we must leave this possibility out of account. In any case it is impossible for us men to compare other people's ideas with our own. I pick the strawberry, I hold it between my fingers. Now my companion sees it too, this same strawberry; but each of us has his own idea. Nobody else has my idea, but many people can see the same thing. Nobody else has my pain. Someone may have sympathy with me, but still my pain belongs to me and his sympathy to him. He has not got my pain, and I have not got his feeling of sympathy.

Fourthly: every idea has only one owner; no two men have the same idea.

For otherwise it would exist independently of this man and independently of that man. Is that lime-tree my idea? By using the expression 'that lime-tree' in this question I am really already anticipating the answer, for I mean to use this expression to designate what I see and other people too can look at and touch. There are now two possibilities. If my intention is realized, if I do designate something with the expression 'that lime-tree', then the thought expressed in the sentence 'That lime-tree is my idea' must obviously be denied. But if my intention is not realized, if I only think I see without really seeing, if on that account the designation 'that lime-tree' is empty, then I have wandered into the realm of fiction without knowing it or meaning to. In that case neither the content of the sentence 'That lime-tree is my idea' nor the content of the sentence 'That lime-tree is not my idea' is true, for in both cases I have a predication which lacks an object. So then I can refuse to answer the question, on the ground that the content of the sentence 'That lime-tree is my idea' is fictional. I have, of course, got an idea then, but that is not what I am using the words 'that lime-tree' to designate. Now someone might really want to designate one of his ideas with the words 'that lime-tree'. He would then be the owner of that which he wants to designate with those words, but then he would not see that lime-tree and no one else would see it or be its owner.

I now return to the question: is a thought an idea? If other people can assent to the thought I express in the Pythagorean theorem just as I do, then it does not belong to the content of my consciousness, I am not its owner; yet I can, nevertheless, acknowledge it as true. However, if what is taken to be the content of the Pythagorean theorem by me and by somebody else is not the same thought at all, we should not really say '*the* Pythagorean theorem' but '*my* Pythagorean theorem', '*his* Pythagorean theorem', and these would be different, for the sense necessarily goes with the sentence. In that case my thought may be the content of my consciousness and his thought the content of his. Could the sense of my Pythagorean theorem be true and the sense of his false? I said that the word 'red' was applicable only in the sphere of my consciousness if it was not meant to state a property of things but to characterize some of my own sense-impressions. Therefore the words 'true' and 'false', as I understand them, might also be applicable only in the realm of my consciousness, if they were not meant to apply to something of which I was not the owner, but to characterize in some way the content of my consciousness. Truth would then be confined to this content and it would remain doubtful whether anything at all similar occurred in the consciousness of others.

If every thought requires an owner and belongs to the contents of his consciousness, then the thought has this owner alone; and there is no science common to many on which many could work, but perhaps I have my science, a totality of thoughts whose owner I am, and another person has his. Each of us is concerned with contents of his own consciousness. No contradiction between the two sciences would then be possible, and it would really be idle to dispute about truth; as idle, indeed almost as ludicrous, as for two people to dispute whether a hundred-mark note were genuine, where each meant the one he himself had in his pocket and understood the word 'genuine' in his own particular sense. If someone takes thoughts to be ideas, what he then accepts as true is, on his own view, the content of consciousness, and does not properly concern other people at all. If he heard from me the opinion that a thought is not an idea he could not dispute it, for, indeed, it would not now concern him.

So the result seems to be: thoughts are neither things in the external world nor ideas.

A third realm must be recognized. Anything belonging to this realm has it in common with ideas that it cannot be perceived by the senses, but has it in common with things that it does not need an owner so as to belong to the contents of his consciousness. Thus, for example, the thought we have expressed in the Pythagorean theorem is timelessly true, true independently of whether anyone takes it to be true. It needs no owner. It is not true only from the time when it is discovered; just as a planet, even before anyone saw it, was in interaction with other planets.[7]

But I think I hear an odd objection. I have assumed several times that the same thing as I see can also be observed by other people. But what if everything were only a dream? If I only dreamt I was walking in the company of somebody else, if I only dreamt that my companion saw the green field as I did, if it were all only a play performed on the stage of my consciousness, it would be doubtful whether there were things of the external world at all. Perhaps the realm of things is empty and I do not see any things or any men, but only have ideas of which I myself am the owner. An idea, being something which can no more exist independently of me than my feeling of fatigue, cannot be a man, cannot look at the same field together with me, cannot see the strawberry I am holding. It is quite incredible that I really have only my inner world, instead of the whole environment in which I supposed myself to move and to act. And yet this is an inevitable consequence of the thesis that only what is my idea can be the object of my awareness. What would follow from this thesis if it were true? Would there then be other men? It would be possible, but I should know nothing of them. For a man cannot be my idea; consequently, if our thesis were true, he cannot be an object of my awareness either. And so this would undercut any reflections in which I assumed that something was an object for somebody else as it was for myself, since even if this were to happen I should know nothing of it. It would be impossible for me to distinguish something owned by myself from something I did not own. In judging something not to be my idea I would make it into the object of my thinking and, therefore, into my idea. On this view, is there a green field? Perhaps, but it would not be visible to me; because if a field is not my idea, it cannot, according to our thesis, be an object of my awareness. But if it is my idea it is invisible, for ideas are not visible. I can indeed have the idea of a green field; however, this is not green, for there are no green ideas. Does a missile weighing a hundred kilogrammes exist, according to this view? Perhaps, but I could know nothing of it. If a missile is not my idea then, according to our thesis, it cannot be an object of my awareness, of my thinking. Yet if a missile were my idea, it would have no weight. I can have an idea of a heavy missile. This then contains the idea of weight as a constituent idea. But this constituent idea is not a property of the whole idea, any more than Germany is a property of Europe. So the consequence is:

Either the thesis that only what is my idea can be the object of my awareness is false, or all my knowledge and perception is limited to the range of my ideas, to the stage of my consciousness. In this case I should have only an inner world and I should know nothing of other people.

It is strange how, in the course of such reflections, opposites turn topsy-turvy. There is, let us suppose, a physiologist of the senses. As is proper for someone investigating nature scientifically, he is at the outset far from supposing the things that he is convinced he sees and touches to be his own ideas. On the contrary, he believes that in sense-impressions he has most reliable evidence of things wholly independent of his feeling, imagining, thinking, which have no need of his consciousness. So little does he consider nerve-fibres and ganglion-cells to be the content of his consciousness that he is, on the contrary, inclined to regard his consciousness as dependent on nerve-fibres and ganglion-cells. He establishes that light-rays, refracted in the eye, strike the visual nerve-endings and there bring about a change, stimulus. From this something is transmitted through nerve-fibres to ganglion-cells. Further processes in the nervous system perhaps follow upon this, and colour-impressions arise, and these perhaps combine to make up what we call the idea of a tree. Physical, chemical, and physiological occurrences get in between the tree and my idea. Only occurrences in my nervous system are immediately connected with my consciousness—or so it seems—and every observer of the tree has his particular occurrences in his particular nervous system. Now light-rays, before they enter my eye, may be reflected by a mirror and diverge as if they came from places behind the mirror. The effects on the visual nerves and all that follows will now take place just as they would if the light-rays had come from a tree behind the mirror and had been propagated undisturbed to the eye. So an idea of a tree will finally occur even though such a tree does not exist at all. The refraction of light, too, with the mediation of the eye and nervous system, may give rise to an idea to which nothing at all corresponds. But the stimulation of the visual nerves need not even happen because of light. If lightning strikes near us, we believe we see flames, even though we cannot see the lightning itself. In this case the visual nerve is perhaps stimulated by electric currents occurring in our body as a result of the flash of lightning. If the visual nerve is stimulated by this means in just the way it would be stimulated by light-rays coming from flames, then we believe we see flames. It just depends on the stimulation of the visual nerve, no matter how that itself comes about.

We can go a step further. Properly speaking this stimulation of the visual nerve is not immediately given; it is only a hypothesis. We believe that a thing independent of us stimulates a nerve and by this means produces a sense-impression; but strictly speaking we experience only that end of the process which impinges on our consciousness. Might not this sense-impression, this sensation, which we attribute to a nerve-stimulation, have other causes also, just as the same nerve-stimulation may arise in different ways? If we call what happens in our consciousness an idea, then we really experience only ideas, not their causes. And if the scientist wants to avoid all mere hypothesis, then he is left just with ideas; everything dissolves into ideas, even the light-rays, nerve-fibres, and ganglion-cells from which he started. So he finally undermines the foundations of his own construction. Is everything an idea? Does everything need an owner without which it can have no existence? I have considered myself as the owner of my ideas, but am I not myself an idea? It seems to me as if I were lying in a deck-chair, as if I could see the toes of a pair of polished boots, the front part of a pair of trousers, the waistcoat, buttons, parts of a jacket, in particular the sleeves, two hands, some hair of a beard, the blurred outline of a nose. Am I myself this entire complex of visual impressions, this aggregate idea? It also seems to me as if I saw a chair over there. That is an idea. I am not actually much different from the chair myself, for am I not myself just a complex of sense-impressions, an idea? But where then is the owner of these ideas? How do I come to pick out one of these ideas and set it up as the owner of the rest? Why need this chosen idea be the idea I like to call 'I'? Could I not just as well choose the one that I am tempted to call a chair? Why, after all, have an owner for ideas at all? An owner would anyhow be something essentially different from ideas that were just owned; something independent, not needing any extraneous owner. If everything is idea, then there is no owner of ideas. And so now once again I experience opposites turning topsy-turvy. If there is no owner of ideas then there are also no ideas, for ideas need an owner and without one they cannot exist. If there is no ruler, there are also no subjects. The dependence which I found myself induced to ascribe to the sensation as contrasted with the sentient being disappears if there no longer is any owner. What I called ideas are then independent objects. No reason remains for granting an exceptional position to that object which I call 'I'.

But is that possible? Can there be an experience without someone to experience it? What would this whole play be without a spectator? Can there be a pain without someone who has it?

Being felt necessarily goes with pain, and, furthermore, someone feeling it necessarily goes with its being felt. But then there *is* something which is not my idea and yet can be the object of my awareness, of my thinking; I myself am such a thing. Or can I be one part of the content of my consciousness, while another part is, perhaps, an idea of the Moon? Does this perhaps take place when I judge that *I* am looking at *the Moon*? Then this first part would have a consciousness, and part of the content of this consciousness would be I myself once more. And so on. Yet it is surely inconceivable that I should be inside myself like this in an infinite nest of boxes, for then there would not be just one *I* but infinitely many. I am not my own idea; and when I assert something about myself, for example that I am not feeling any pain at the moment, then my judgement concerns something which is not a content of my consciousness, is not my idea, namely myself. Therefore that about which I state something is not necessarily my idea. But someone perhaps objects: if I think I have no pain at the moment, does not the word 'I' answer to something in the content of my consciousness? And is that not an idea? That may be so. A certain idea in my consciousness may be associated with the idea of the word 'I'. But then this is one idea among other ideas, and I am its owner as I am the owner of the other ideas. I have an idea of myself, but I am not identical with this idea. What is a content of my consciousness, my idea, should be sharply distinguished from what is an object of my thought. Therefore the thesis that only what belongs to the content of my consciousness can be the object of my awareness, of my thought, is false.

Now the way is clear for me to acknowledge another man likewise as an independent owner of ideas. I have an idea of him, but I do not confuse it with him himself. And if I state something about my brother, I do not state it about the idea that I have of my brother.

The patient who has a pain is the owner of this pain, but the doctor who is treating him and reflects on the cause of this pain is not the owner of the pain. He does not imagine he can relieve the pain by anaesthetizing himself. There may very well be an idea in the doctor's mind that answers to the patient's pain, but that is not the pain, and is not what the doctor is trying to remove. The doctor might consult another doctor. Then one must distinguish: first, the pain, whose owner is the patient; secondly, the first doctor's idea of this pain; thirdly, the second doctor's idea of this pain. This last idea does indeed belong to the content of the second doctor's consciousness, but it is not the object of his reflection; it is, rather, an aid to reflection, as a drawing may be. The two doctors have as their common objective of thought the patient's pain, which they do not own. It may be seen from this that not only a thing but also an idea may be a common object of thought for people who do not have the idea.

In this way, it seems to me, the matter becomes intelligible. If man could not think and could not take as the object of his thought something of which he was not the owner, he would have an inner world but no environment. But may this not be based on a mistake? I am convinced that the idea I associate with the words 'my brother' corresponds to something that is not my idea and about which I can say something. But may I not be making a mistake about this? Such mistakes do happen. We then, against our will, lapse into fiction. Yes, indeed! By the step with which I win an environment for myself I expose myself to the risk of error. And here I come up against a further difference between my inner world and the external world. I cannot doubt that I have a visual impression of green, but it is not so certain that I see a lime-leaf. So, contrary to widespread views, we find certainty in the inner world, while doubt never altogether leaves us in our excursions into the external world. But the probability is nevertheless in many cases hard to distinguish from certainty, so we can venture to judge about things in the external world. And we must make this venture even at the risk of error if we do not want to fall into far greater dangers.

As the result of these last considerations I lay down the following: not everything that can be the object of my acquaintance is an idea. I, being owner of ideas, am not myself an idea. Nothing now stops me from acknowledging other men to be owners of ideas, just as I am myself. And, once given the possibility, the probability is very great, so great that it is in my opinion no longer distinguishable from certainty. Would there be a science of history otherwise? Would not all moral theory, all law, otherwise collapse? What would be left of religion? The natural sciences too could only be assessed as fables like astrology and alchemy. Thus the reflections I have set forth on the assumption that there are other men, besides myself, who can make the same thing the object of their consideration, their thinking, remain in force without any essential weakening.

Not everything is an idea. Thus I can also acknowledge thoughts as independent of me; other men can grasp them just as much as I; I can acknowledge a science in which many can be engaged in research. We are not owners of thoughts as we are owners of our ideas. We do not *have* a thought as we have, say, a sense-impression, but we also do not *see* a thought as we see, say, a star. So it is advisable to choose a special expression; the word 'grasp' suggests itself for the purpose.[8] To the grasping of thoughts there must then correspond a special mental capacity, the power of thinking. In thinking we do not produce thoughts, we grasp them. For what I have called thoughts stand in the closest connection with truth. What I acknowledge as true, I judge to be true quite apart from my acknowledging its truth or even thinking about it. That someone thinks it has nothing to do with the truth of a thought. 'Facts, facts, facts' cries the scientist if he wants to bring home the necessity of a firm foundation for science. What is a fact? A fact is a thought that is true. But the scientist will surely not acknowledge something to be the firm foundation of science if it depends on men's varying states of consciousness. The work of science does not consist in creation, but in the discovery of true thoughts. The astronomer can apply a mathematical truth in the investigation of long-past events which took place when—on Earth at least—no one had yet recognized that truth. He can do this because the truth of a thought is timeless. Therefore that truth cannot have come to be only upon its discovery.

Not everything is an idea. Otherwise psychology would contain all the sciences within it, or at least it would be the supreme judge over all the sciences. Otherwise psychology would rule even over logic and mathematics. But nothing would be a greater misunderstanding of mathematics than making it subordinate to psychology. Neither logic nor mathematics has the task of investigating minds and the contents of consciousness owned by individual men. Their task could perhaps be represented rather as the investigation of *the* mind; of *the* mind, not of minds.

The grasp of a thought presupposes someone who grasps it, who thinks. He is the owner of the thinking, not of the thought. Although the thought does not belong with the contents of the thinker's consciousness, there must be something in his consciousness that is aimed at the thought. But this should not be confused with the thought itself. Similarly, Algol itself is different from the idea someone has of Algol.

A thought belongs neither to my inner world as an idea, nor yet to the external world, the world of things perceptible by the senses.

This consequence, however cogently it may follow from the exposition, will nevertheless perhaps not be accepted without opposition. It will, I think, seem impossible to some people to obtain information about something not belonging to the inner world except by sense-perception. Sense-perception indeed is often thought to be the most certain, even the sole, source of knowledge about everything that does not belong to the inner world. But with what right? For sense-perception has as necessary constituents our sense-impressions and these are a part of the inner world. In any case two men do not have the same sense-impressions, though they may have similar ones. Sense-impressions alone do not reveal the external world to us. Perhaps there is a being that has only sense-impressions without seeing or touching things. To have visual impressions is not to see things. How does it happen that I see the tree just there where I do see it? Obviously it depends on the visual impressions I have and on the particular sort which occur because I see with two eyes. On each of the two retinas there arises, physically speaking, a particular image. Someone else sees the tree in the same place. He also has two retinal images but they differ from mine. We must assume that these retinal images determine our impressions. Consequently, the visual impressions we have are not only not the same, but markedly different from each other. And yet we move about in the same external world. Having visual impressions is certainly necessary for seeing things, but not sufficient. What must still be added is not anything sensible. And yet this is just what opens up the external world for us; for without this non-sensible something everyone would remain shut up in his inner world. So perhaps, since the decisive factor lies in the non-sensible, something non-sensible, even without the co-operation of sense-impressions, could also lead us out of the inner world and enable us to grasp thoughts. Outside our inner world we should have to distinguish the external world proper of sensible, perceptible things and the realm of what is non-sensibly perceptible. We should need something non-sensible for the recognition of both realms; but for the sense-perception of things we should need sense-impressions as well, and these belong entirely to the inner world. So the distinction between the ways in which a thing and a thought are given mainly consists in something which is assignable, not to either of the two realms, but to the inner world. Thus I cannot find this distinction to be so great as to make impossible the presentation of a thought that does not belong to the inner world.

A thought, admittedly, is not the sort of thing to which it is usual to apply the term 'actual'. The world of actuality is a world in which this acts on that and changes it and again undergoes reactions itself and is changed by them. All this is a process in time. We will hardly admit what is timeless and unchangeable to be actual. Now is a thought changeable or is it timeless? The thought we express by the Pythagorean theorem is surely timeless, eternal, unvarying. But are there not thoughts which are true today but false in six months time? The thought, for example, that the tree there is covered with green leaves will surely be false in six months' time. No, for it is not the same thought at all. The words 'This tree is covered with green leaves' are not sufficient by themselves to constitute the expression of thought, for the time of utterance is involved as well. Without the time-specification thus given we have not a complete thought, that is we have no thought at all. Only a sentence with the time-specification filled out, a sentence complete in every respect, expresses a thought. But this thought, if it is true, is true not only today or tomorrow but timelessly. Thus the present tense in 'is true' does not refer to the speaker's present; it is, if the expression be permitted, a tense of timelessness. If we merely use the assertoric sentence-form and avoid the word 'true', two things must be distinguished, the expression of the thought and assertion. The time-specification that may be contained in the sentence belongs only to the expression of the thought; the truth, which we acknowledge by using the assertoric sentence-form, is timeless. Of course, the same words, on account of the variability of language with time, may take on another sense, express another thought; this change, however, relates only to the linguistic realm.

And yet what value could there be for us in the eternally unchangeable, which could neither be acted upon nor act on us? Something entirely and in every respect inactive would be quite unactual, and so far as we are concerned it would not be there. Even the timeless, if it is to be anything for us, must somehow be implicated with the temporal. What would a thought be for me if it were never grasped by me? But by grasping a thought I come into a relation to it, and it to me. It is possible that the same thought as is thought by me today was not thought by me yesterday. Of course this does away with strict timelessness. But we may be inclined to distinguish between essential and inessential properties and to regard something as timeless if the changes it undergoes involve only inessential properties. A property of a thought will be called inessential if it consists in, or follows from, the fact that this thought is grasped by a thinker.

How does a thought act? By being grasped and taken to be true. This is a process in the inner world of a thinker which may have further consequences in this inner world, and which may also encroach on the sphere of the will and make itself noticeable in the outer world as well. If, for example, I grasp the thought we express by the theorem of Pythagoras, the consequence may be that I recognize it to be true, and further that I apply it in making a decision which brings about the acceleration of masses. This is how our actions are usually led up to by acts of thinking and judging. And so thoughts may indirectly influence the motion of masses. The influence of man on man is brought about for the most part by thoughts. People communicate thoughts. How do they do this? They bring about changes in the common external world, and these are meant to be perceived by someone else, and so give him a chance to grasp a thought and take it to be true. Could the great events of world history have come about without the communication of thoughts? And yet we are inclined to regard thoughts as unactual, because they appear to do nothing in relation to events, whereas thinking, judging, stating, understanding, in general doing things, are affairs that concern men. How very different the actuality of a hammer appears, compared with that of a thought! How different a process handing over a hammer is from communicating a thought! The hammer passes from one control to another, it is gripped, it undergoes pressure, and thus its density, the disposition of its parts, is locally changed. There is nothing of all this with a thought. It does not leave the control of the communicator by being communicated, for after all man has no power over it. When a thought is grasped, it at first only brings about changes in the inner world of the one who grasps it; yet it remains untouched in the core of its essence, for the changes it undergoes affect only inessential properties. There is lacking here something we observe everywhere in physical process—reciprocal action. Thoughts are not wholly unactual but their actuality is quite different from the actuality of things. And their action is brought about by a performance of a thinker;

without this they would be inactive, at least as far as we can see. And yet the thinker does not create them but must take them as they are. They can be true without being grasped by a thinker; and they are not wholly unactual even then, at least if they *could* be grasped and so brought into action.

Notes

1. So, similarly, people have said 'a judgement is something which is either true or false'. In fact I use the word 'thought' more or less in the sense 'judgement' has in the writings of logicians. I hope it will become clear in the sequel why I choose 'thought'. Such an explanation has been objected to on the ground that it makes a division of judgements into true and false judgements—perhaps the least significant of all possible divisions among judgements. But I cannot see that it is a logical fault that a division is given along with the explanation. As for the division's being significant, we shall perhaps find we must hold it in no small esteem, if, as I have said, it is the word 'true' that points the way for logic.
2. I am not using the word 'sentence' here in quite the same sense as grammar does, which also includes subordinate clauses. An isolated subordinate clause does not always have a sense about which the question of truth can arise, whereas the complex sentence to which it belongs has such a sense.
3. Frege means a question introduced by an interrogative word like 'who?' (trans.)
4. i.e. yes–no questions: German *Satzfragen*. (trans.)
5. It seems to me that thought and judgement have not hitherto been adequately distinguished. Perhaps language is misleading. For we have no particular bit of assertoric sentences which corresponds to assertion; that something is being asserted is implicit rather in the assertoric form. We have the advantage in German that main and subordinate clauses are distinguished by the word order. However, in this connection we must observe that a subordinate clause may also contain an assertion, and that often neither main nor subordinate clause expresses a complete thought by itself but only the complex sentence does.
6. I am not here in the happy position of a mineralogist who shows his audience a rock-crystal: I cannot put a thought in the hands of my readers with the request that they should examine it from all sides. Something in itself not perceptible by sense, the thought, is presented to the reader—and I must be content with that—wrapped up in a perceptible linguistic form. The pictorial aspect of language presents difficulties. The sensible always breaks in and makes expressions pictorial and so improper. So one fights against language, and I am compelled to occupy myself with language although it is not my proper concern here. I hope I have succeeded in making clear to my readers what I mean by 'a thought'.
7. A person sees a thing, has an idea, grasps or thinks a thought. When he grasps or thinks a thought he does not create it but only comes to stand in a certain relation to what already existed—a different relation from seeing a thing or having an idea.
8. The expression 'grasp' is as metaphorical as 'content of consciousness'. The nature of language does not permit anything else. What I hold in my hand can certainly be regarded as the content of my hand; but all the same it is the content of my hand in quite another and a more extraneous way than are the bones and muscles of which the hand consists or, again, the tensions these undergo.

III

SELECTION FROM THE FREGE–RUSSELL CORRESPONDENCE*

I. EXCERPT FROM FREGE TO RUSSELL, 13 NOVEMBER 1904

Jena
13 November 1904

Dear Colleague,

. . . Mont Blanc with its snowfields is not itself a component part of the thought that Mont Blanc is more than 4,000 metres high. . . . The sense of the word 'moon' is a component part of the thought that the moon is smaller than the earth. The moon itself (i.e. the denotation of the word 'moon') is not part of the sense of the word 'moon'; for then it would also be a component part of that thought. We can nevertheless say: 'The moon is identical with the heavenly body closest to the earth'. What is identical, however, is not a component part but the denotation of the expressions 'the moon' and 'the heavenly body closest to the earth'. We can say that 3 + 4 is identical with 8 − 1; i.e. that the denotation of '3 + 4' coincides with the denotation of '8 − 1'. But this denotation, namely the number 7, is not a component part of the sense of '3 + 4'. The identity is not an identity of sense, nor of part of the sense, but of denotation. . . .

Yours sincerely
G. Frege

* The translation is Hans Kaal's, except that *Bedeutung* is here translated as 'denotation'.

Selection from the Frege–Russell correspondence in Gottlob Frege, *Philosophical and Mathematical Correspondence*, ed. Gabriel, Hermes, Kambartek, Thiel, and Veraart (University of Chicago Press, 1980), pp. 163, 169.

II. EXCERPT FROM RUSSELL TO FREGE, 12 DECEMBER 1904

Ivy Lodge
Tilford, Farnham
12 December 1904

Dear Colleague,

. . . Concerning sense and denotation, I see nothing but difficulties which I cannot overcome. . . . I believe that in spite of all its snowfields Mont Blanc itself is a component part of what is actually asserted in the proposition 'Mont Blanc is more than 4,000 metres high'. We do not assert the thought, for this is a private psychological matter: we assert the object of the thought, and this is, to my mind, a certain complex (an objective proposition, one might say) in which Mont Blanc is itself a component part. If we do not admit this, then we get the conclusion that we know nothing at all about Mont Blanc. This is why for me the *denotation* of a proposition is not the true, but a certain complex which (in the given case) is true. In the case of a simple proper name like 'Socrates', I cannot distinguish between sense and denotation; I see only the idea, which is psychological, and the object. Or better: I do not admit the sense at all, but only the idea and the denotation. I see the difference between sense and denotation only in the case of complexes whose denotation is an object, e.g. the values of ordinary mathematical functions like $\xi + 1$, ξ^2, etc. . . .

Yours sincerely
BERTRAND RUSSELL

IV

A REMARK CONCERNING QUINE'S PARADOX ABOUT MODALITY

ALONZO CHURCH

IN what at first sight may seem to be its simplest and most direct form, Quine's paradox about modality may be explained by reference to the following example:

(1) $\Box\, 9 = 9$

(2) The number of major planets $= 9$

Here '$\Box$' is the notation for necessity in standard modal logic and may be read as 'it is necessary that'. Both (1) and (2) are commonly accepted as true, the former on logico–mathematical grounds and the latter on astronomical grounds. The paradox arises if, relying on the principle of substitutivity of equality, we make use of (2) in order to substitute 'the number of major planets' for the first occurrence of '9' in (1). For in this way we infer from two accepted truths what is evidently false:

(3) $\Box$ the number of major planets $= 9$

Indeed it was at one time believed that the number of major planets is less than 9; and this belief, though factually false, was hardly a belief in an impossibility.

Quine's original formulation of the paradox does not use exactly this example but a number of others—which, however, illustrate what is evidently the same logical difficulty.[1] And one of Quine's examples in 'Reference and Modality' is in fact the same as the above except that the premiss (1) is replaced by

(4) $\Box\, 9 > 7$,

Alonzo Church, 'A Remark Concerning Quine's Paradox About Modality', Spanish version in *Analisis Filosófico* 2 (1982), 25–32.

so that, in place of (3), the paradoxical conclusion becomes:

(5) $\Box$ the number of major planets > 7

The inference of (3) from the premisses (1) and (2) has here been chosen as an example in order to exhibit more clearly the very close parallelism between Quine's paradox and the paradox which Russell[2] sought to eliminate by means of his theory of descriptions.

For the latter paradox we may use a minor variation of Russell's example about King George IV.[3] We may assume it true, as of some appropriate date, that

(6) George IV believes that Sir Walter Scott = Sir Walter Scott,

since George IV, being well acquainted with Sir Walter, is unlikely to doubt this. Moreover, it is factually true, as of this same date, that

(7) The author of *Waverley* = Sir Walter Scott.

By substitutivity of equality it seems to follow that

(8) George IV believes that the author of *Waverley* = Sir Walter Scott.

But (8) is known to be false, still as of the same date; and even without the factual information it is clearly unreasonable to suppose that (8) is a logical consequence of (6) and (7).

Or still better we may use the premiss

(9) George IV believes that 9 = 9,

which is at least probably true. From (9) and (2) there seems to follow by substitutivity of equality:

(10) George IV believes that the number of major planets = 9.

But (10) is certainly false, because the discovery of the last two major planets came only after King George's death.

Given the truth of (2), the paradox based on the inference of (3) from (1) differs from that based on the inference of (10) from (9) only in the replacement of 'George IV believes that' by the modal operator ' $\Box$ '. It may therefore be argued that there are not two genuinely different paradoxes, but only various examples illustrating what must be regarded as a single paradox. And indeed Quine, already in 'Notes on Existence and Necessity', makes a close association between paradoxes about belief and paradoxes about modality.

This paradox may arise not only in connection with 'believes that' and 'it is necessary that' but also with any of various other phrases which we may speak of as introducing intensional contexts. Carnap calls it the *antinomy of the name relation*.[4] And this terminology is useful as a reminder that, in place of substitutivity of equality, the paradox may be made to depend on the semantical principle that (in Carnap's words) *if two expressions name the same entity, then a true sentence remains true when the one is replaced in it by the other*. But since—contrary to Carnap—there is no actual antinomy or contradiction, but only such results as (3), (8), (10), which are factually false or unacceptably counterintuitive, the writer prefers to speak of the *paradox of the name relation*.

Once it is seen that Quine's paradox about modality and the paradox about King George IV and Sir Walter Scott must be seen as instances of the same paradox, it is not surprising that Smullyan is able to resolve Quine's paradox by means of Russell's theory of descriptions.[5] Moreover, it seems certain in advance that whatever objections may apply to Smullyan's modal logic with descriptions (e.g. Quine's objection in 'Reference and Modality' that it requires excessive attention to matters of scope) must equally apply to the logic of belief statements with descriptions which is implicit in Russell's resolution in 'On Denoting' of the paradox about George IV and Sir Walter.[6]

Besides the complications regarding scope which arise when the paradox about modality is resolved by Russell's doctrine of descriptions, Quine's 'Reference and Modality' raises also a different objection, which depends on reformulating the paradox in a way that refers only to variables and makes no use either of names (or naming expressions) or of desriptions. The point is that although Russell largely reconstrues names and naming expressions as descriptions, and then eliminates the descriptions by his device of contextual definition,[7] of course the use of variables is not thereby eliminated.

Citing Ruth Barcan[8] Quine calls attention to the theorem

(11) $x = y \supset_{xy} \Box\, x = y.$

Indeed this theorem follows by elementary logic alone, independently of the exact meaning or definition of the sign '=' of identity or equality, provided only that we have

(12) $x = y \supset_{xy} . F(x) \supset F(y)$

and

(13) $(x) \Box x = x.$

For by substitution in (12) we get

(14) $x = y \supset_{xy} . \Box[x = x] \supset \Box[x = y].$

And (11) then follows by (13) and (14).

The result is perhaps more striking when put in terms of possibility rather than necessity. For

(15) $\sim F(x) \supset_x . F(y) \supset_y x \neq y$

is equally an elementary property of identity which can hardly be denied. And if we allow further that

(16) $(x) \sim\Diamond x \neq x,$

we obtain by substitution in (15) that

(17) $\sim\Diamond [x \neq x] \supset_{x} . \Diamond [x \neq y] \supset_y x \neq y.$

Then from (16) and (17) we have the following variant of Murphy's Law:

(18) $\Diamond [x \neq y] \supset_{xy} x \neq y.$

(If two things are possibly different, then they *are* different.)

The theorems (11) and (18) are hardly avoidable if modal logic is formulated in such a way that modal operators are prefixed directly to sentences (as indeed is now usual).[9] Quine in 'Reference and Modality' objects to these theorems as compelling acceptance of 'Aristotelian essentialism'[10]—which he regards as philosophically suspect, and as being, moreover, incompatible with the idea which many modal logicians have held, that the modal sentence $\Box S$ is true if and only if the corresponding unmodalized sentence S is analytic.

This second form of Quine's paradox about modality, which refers to variables rather than names,[11] can be paralleled by a paradox about belief statements in place of modality. By substitution in (15) we get:

(19) For every x and every y, if George IV does not believe that $x \neq x$, if George IV believes that $x \neq y$, then $x \neq y$.

This may be thought of as analogous to (17). The second premiss,

(20) For every x, George IV does not believe that $x \neq x$,

differs from (16) in being no more than very likely rather than certain. But if we accept it, we have as analogous to (18) the conclusion:

(21) For every x and every y, if George IV believes that $x \neq y$, then $x \neq y$.

And this otherwise surprising power of King George's beliefs to control the actual facts about x and y can be explained only on the doubtful assumption that belief properly applies 'to the fulfilment of conditions by objects' quite 'apart from special ways of specifying' the objects.[12] This assumption (let us call it the principle of transparency of belief) is the same thing to the notion of belief that essentialism is to the notion of necessity. But the consequences of the former for the ordinary notion of belief may be thought to be even more repellent than the consequences of essentialism for modal notions.

To illustrate this last point, let us suppose that George IV is convinced (and in fact on good evidence) that there is one and only one who wrote *Waverley*, that there is one and only one who wrote *Ivanhoe*, and that the two authors are the same. Then let us ask for what objects[13] (or individuals) y we have that

(22) George IV believes that y wrote *Waverley*.

The ordinary notion of belief seems to require that although (22) holds when y is specified in a special way, namely as having written *Ivanhoe*, it may yet fail when the same y is specified in some other special way, for example as scottizing.

Our conclusion is that Quine's objections against the Russellian treatment[14] of modal logic, according to which modal operators are prefixed to sentences, do have some considerable force. But it is better to present them in a way that exhibits the nearly complete parallelism between the objections against a Russellian modal logic and those against a Russellian logic of belief (or of denying, wishing to know, or the like). For this has the effect of putting the objections in perspective and of clarifying both their strengths and their weaknesses.

In summary, the significant objections in 'Reference and Modality' are two: the complications about scope which arise in connection with the use of descriptions, and the transparency of both belief and necessity which is forced by use of the theory of descriptions to resolve the paradox of the name relation.

But finally it must be pointed out that Quine's objections, though strong, are no firm refutation of the Russellian resolution of the paradox. There may be those who, in the interest of the resolution of the paradox, are willing to accept both the complications about descriptions and the strange transparent notions of belief and necessity which result. And to them it can only be said that well, it does seem strange. The sort of essentialism to which the transparency of notions such as belief and necessity leads does not rise above the level of variables and primitive constants. And the Russellian may proceed with the more confidence because he is able, besides the transparent notions of belief, necessity, and possibility, to express also the more usual non-transparent notions. Namely if $B(x, p)$ is used to mean that x believes that p,[15] $S(x)$ that x scottizes, $W(x)$ that x is author of *Waverley*, $G(x)$ that x is George IV (or that x georgivizes), he may write:[16]

(23) $B(\ (\iota x)G(x), (\iota x)S(x) = (\iota x)W(x)\ ).$

(24) $\Diamond\ (\iota x)S(x) \neq (\iota x)W(x).$

The convention of minimum scope is of course to be understood in (23) and (24). And notwithstanding (18), it does not follow from (24) that

$$(\iota x)S(x) \neq (\iota x)W(x).$$

Notes

1. W. V. Quine, 'Notes on Existence and Necessity', *Journal of Philosophy* 40 (1943), 113–27 (reprinted in Leonard Linsky (ed.), *Semantics and the Philosophy of Language* (Urbana, 1952), 77–91) and 'Reference and Modality' in *From a Logical Point of View* (Cambridge, Mass., 1961), 139–59 (reprinted in Leonard Linsky (ed.), *Reference and Modality* (London, 1971), 17–34.
2. Bertrand Russell, 'On Denoting', *Mind* 14 (1905), 479–93. Reprinted in Herbert Feigl and Wilfred Sellars (eds.), *Readings in Philosophical Analysis* (New York, 1949), 103–15; Robert C. Marsh (ed.), *Logic and Knowledge, Essays 1901–1950, by Bertand Russell* (London, 1956), 41–56; Irving M. Copi and James A. Gould (eds.), *Contemporary Readings in Logical Theory* (New

York and London, 1967), 93–105; Ausanis Marras (ed.), *Intentionality, Mind, and Language* (Urbana, Chicago, and London, 1972), 362–79.

3. The example is due to Russell, 'On Denoting'; the point which it illustrates, to Gottlob Frege, 'Uber Sinn und Bedeutung', *Zeitschrift für Philosophie und Philosophische Kritik* 100 (1892), 25–50. (Reprinted in Gunther Panzig (ed.), *Funktion, Begriff, Bedeutung: Fünf logische Studien*[2] (Gottingen, 1966), 40–65. English translation in Feigl–Sellars, pp. 85–102, reprinted in Copi–Gould, pp. 75–92, and in Marras, pp. 337–61.)
4. Rudolf Carnap, *Meaning and Necessity*[2] (Chicago, 1956).
5. Arthur F. Smullyan, Review of Quine's 'The Problem of Interpreting Modal Logic' (*Journal of Symbolic Logic* 12 (1947), 43–8; reprinted in Copi–Gould, pp. 267–73) in *Journal of Symbolic Logic* 12 (1947), 139–41 and 'Modality and Description', *Journal of Symbolic Logic* 13 (1948), 31–7 (reprinted in Linsky, *Reference and Modality*, pp. 35–43).
6. Russell has an informal treatment of the matter of scope of descriptions in connection with belief statements (and the like), using the particular examples 'I thought your yacht was larger than it is' and 'George IV wished to know whether Scott was the author of *Waverley*'; but in his later writings, leading up to and including *Principia Mathematica*, (A. N. Whitehead and Bertrand Russell, *Principia Mathematica*, 3 vols. (Cambridge, 1910–13)), he does not return to this. His formal language is confined to what is needed for the *Principia* account of the foundations of mathematics, and he therefore never considers a formalized logic of belief in connection with the theory of descriptions. Nevertheless, the complications about scope are already implicit in Russell's paper of 1905 and are not due to changes by Smullyan.
7. As Russell writes in 'On Denoting': 'The phrase *per se* has no meaning, because in any proposition in which it occurs the proposition, fully expressed, does not contain the phrase, which has been broken up.
8. Ruth C. Barcan, 'The Identity of Individuals in a Strict Functional Calculus of Second Order', *Journal of Symbolic Logic* 12 (1947), 12–15, and her earlier papers.
9. The writer believes that a more Fregean version of modal logic might be preferable, in which the modal operators are prefixed not directly to a sentence but to any name of the proposition which the sentence expresses. Quine's misgivings about this in 'Reference and Modality' can be dispelled only by a detailed development of such a Fregean modal logic, explicitly exhibiting the 'interplay' which he fears may be wanting; but Quine's further objection (in W. V. Quine, *Word and Object* (Cambridge, Mass., 1960); cf. printing of 1973, p. 198) that there is some *ad hoc* restriction on quantifying into modal contexts seems to be based on a misunderstanding. (For historical accuracy it should be added that Frege himself disbelieved in modal logic.)
10. Cf. also *Word and Object*, § 41.
11. Or naming expressions. No distinction is intended in this paper, or by Carnap, between names and naming expressions. And the idea is Russellian rather than Fregean, that a name must be an unanalysed primitive and hence normally a single word or a single symbol.
12. The quoted phrases are from Section III of 'Reference and Modality' and are used in order to emphasize that what is here said about belief closely parallels what is said by Quine about modality.
13. We assume that human beings are included among objects, or among

individuals (as it may be better to say in order to allow for type theory and its standard terminology). And we follow Quine in avoiding the semantical formulation that consists in asking what values of the variable 'y' satisfy the propositional form 'George IV believes that y wrote *Waverley*'—although this alternative formulation might otherwise be helpful, e.g. in bringing out that what is at issue concerns the values of a variable, and not (as in the original paradox of the name relation) the denotation of a name or names.

14. We call it Russellian in spite of Russell's own rejection of modality, because it is the treatment appropriate to Russell's explicit and implicit semantics, especially propositions as values of the propositional variables, and the sort of transparency in belief contexts, modal contexts, and the like that is required by the theory of descriptions as resolution of the paradox of the name relation.
15. To avoid antinomies such as the *Epimenides*, it may be necessary either to distinguish different orders of propositional variables (by adopting ramified type theory) or to distinguish different orders of belief by writing B^1, B^2, B^3, etc. We ignore this here as not being immediately relevant to what is being said.
16. These examples illustrate that Russell's unitary propositional functions, where e.g. 'F is unitary', is defined as $(\exists x).\ F(y) \equiv_y y = x$, may serve to a considerable extent as surrogates for the entities which in Frege's theory appear as senses of names. For example in (23) King George's non-transparent belief appears as a relation between the propositional functions S and W (or between $S(\hat{x})$ and $W(\hat{x})$ as Russell would write), and these propositional functions must be unitary if King George's belief is not mistaken. And in a Fregean theory the same belief by George IV would appear as a relation between the senses which belong to the names 'Sir Walter Scott' and 'the author of *Waverley*'. It would be of interest to look into the question how far the Fregean theory can be reproduced within the Russellian by identifying the Fregean senses with propositional functions (in Russell's sense, according to which propositional functions are intensional entities). Indeed, the Russellian theory might have to be rather drastically mutilated to obtain the Fregean fragment. But it remains true that any significant partial success in representing one theory within the other would throw light on the relationship of the two theories.

V

ON THE LOGIC OF DEMONSTRATIVES*

DAVID KAPLAN

In this paper, I propose to outline briefly a few results of my investigations into the theory of demonstratives: words and phrases whose *in*tension is determined by the contexts of their use. Familiar examples of demonstratives are the nouns 'I', 'you', 'here', 'now', 'that', and the adjectives 'actual' and 'present'. It is, of course, clear that the *ex*tension of 'I' is determined by the context—if you and I both say 'I' we refer to different persons. But I would now claim that the intension is also so determined. The intension of an 'eternal' term (like 'The Queen of England in 1973') has generally been taken to be represented by a function which assigns to each possible world the Queen of England in 1973 of that world. Such functions would have been called *individual concepts* by Carnap. It has been thought by some—myself among others—that by analogy, the intension of 'I' could be represented by a function from speakers to individuals (in fact, the identity function). And, similarly, that the intensions of 'here' and 'now' would be represented by (identity) functions on places and times. The role of contextual factors in determining the extension (with respect to such factors) of a demonstrative was thought of as analogous to that of a possible world in determining the extension of 'the Queen of England in 1973' (with respect to that possible world). Thus an enlarged view of an intension was derived. The intension of an expression was to be represented by a function from certain factors to the extension of an expression (with respect to those factors). Originally such factors were simply possible worlds, but as it was noticed that the so-called tense operators exhibited a structure highly analogous to that of the modal operators, the factors with respect to which extension was to be

David Kaplan, 'On The Logic of Demonstratives', *Journal of Philosophical Logic* 8 (1978), 81–98.

determined were enlarged to include moments of time. When it was noticed that contextual factors were required to determine the extension of sentences containing demonstratives, a still more general notion was developed and called an 'index'. The extension of an expression was to be determined with respect to an index. The intension of an expression was that function which assigned to every index the extension at that index. Here is a typical passage.

The above example supplies us with a statement whose truth-value is not constant but varies as a function of $i \in I$. This situation is easily appreciated in the context of time-dependent statements; that is, in the case where I represents the instants of time. Obviously the same statement can be true at one moment and false at another. For more general situations one must not think of the $i \in I$ as anything as simple as instants of time or even possible worlds. In general we will have

$$i = (w, t, p, a, \ldots)$$

where the index i has many *coordinates*: for example, w is a *world*, t is a *time*, $p = (x, y, z)$ is a (3-dimensional) *position* in the world, a is an *agent*, etc. All these coordinates can be varied, possibly independently, and thus affect the truth values of statements which have indirect reference to these coordinates. (From the Advice of a prominent logician.)

A sentence ϕ was taken to be logically true if true at every index (in every 'structure'), and $\Box\,\phi$ was taken to be true at a given index (in a given structure) just in case ϕ was true at every index (in that structure). (Or possibly, just in case ϕ was true at every index *which differed from the given index only in possible world coordinate*.) Thus the familiar principle of modal generalization: if $\models \phi$, then $\models \Box\,\phi$, is validated.

This view, in its treatment of demonstratives, now seems to me to have been technically wrong (though perhaps correctable by minor modification) and, more importantly, conceptually misguided.

Consider the sentence

(1) I am here now.

It is obvious that for many choices of index—i.e. for many quadruples $\langle w, x, p, \mathrm{t}\rangle$ where w is a possible world, x is a person, p is a place, and t is a time—(1) will be false. In fact, (1) is true only with respect to those indices $\langle w, x, p, \mathrm{t}\rangle$ which are such that in the world w, x is located at p at the time t. Thus (1) fares about on a par with

(2) David Kaplan is in Los Angeles on 21 April 1973.

(2) is contingent, and so is (1).

But here we have missed something essential to our understanding of demonstratives. Intuitively, (1) is deeply, and in some sense universally, true. One need only understand the meaning of (1) to know that it cannot be uttered falsely. No such guarantees apply to (2). A *Logic of Demonstratives* which does not reflect this intuitive difference between (1) and (2) has bypassed something essential to the logic of demonstratives.

Here is a proposed correction. Let the class of indices be narrowed to include only the *proper* ones—namely, those $\langle w, x, p, \mathrm{t}\rangle$ such that in the world *w*, *x is* located at *p* at the time *t*. Such a move may have been intended originally since improper indices are like impossible worlds; no such contexts *could* exist and thus there is no interest in evaluating the extensions of expressions with respect to them. Our reform has the consequence that (1) comes out, correctly, to be logically true. Now consider

(3) □ I am here now.

Since the contained sentence (namely (1)) is true at every proper index, (3) also is true at every proper index and thus also is logically true. (As would be expected by the aforementioned principle of modal generalization.)

But (3) should not be *logically* true, since it is false. It is certainly *not* necessary that I be here now. But for several contingencies I would be working in my garden now, or even writing this in a location outside of Los Angeles.

Perhaps enough has now been said to indicate that there are difficulties in the attempt to assimilate the role of a *context* in a logic of demonstratives to that of a *possible world* in the familiar modal logics or a *moment of time* in the familiar tense logics.

I believe that the source of the difficulty lies in a conceptual confusion between two kinds of meaning. Ramifying Frege's distinction between sense and denotation, I would add two varieties of sense: content and character. The content of an expression is always taken *with respect to* a given context of use. Thus when I say

(4) I was insulted yesterday

a specific content—*what I said*—is expressed. Your utterance of the same sentence, or mine on another day, would not express the same content. What is important to note is that it is not just the truth value that may change; what is said is itself different. Speaking today, my utterance of (4) will have a content roughly equivalent to that which

(5) David Kaplan is insulted on 20 April 1973

would have spoken by you or anyone at any time. Since (5) contains no demonstratives, its content is the same with respect to all contexts. This content is what Carnap called an 'intension' and what, I believe, has been often referred to as a 'proposition'. So my theory is that different contexts for (4) produce not just different truth values, but different propositions.

Turning now to character, I call that component of the sense of an expression which determines how the content is determined by the context the 'character' of an expression. Just as contents (or intensions) can be represented by functions from possible worlds to extensions, so characters can be represented by functions from contexts to contents. The character of 'I' would then be represented by *the function (or rule, if you prefer) which assigns to each context that content which is represented by the constant function from possible words to the agent of the context*. The latter function has been called an 'individual concept'. Note that the character of 'I' is represented by a function from contexts to individual *concepts*, not from contexts to individuals. It was the idea that a function from contexts to individuals could represent the intension of 'I' which led to the difficulties discussed earlier.

Now what is it that a competent speaker of English knows about the world 'I'? Is it the content with respect to some particular occasion of use? No. It is the character of 'I': the rule italicized above. Competent speakers recognize that the proper use of 'I' is —loosely speaking—to refer to the speaker. Thus, that component of sense which I call 'character' is best identified with what might naturally be called 'meaning'.

To return, for a moment, to (1). The character (meaning) of (1) determines each of the following:

(*a*) In different contexts, an utterance of (1) expresses different contents (propositions).

(*b*) In most (if not all) contexts, an utterance of (1) expresses a contingent proposition.

(*c*) In all contexts, an utterance of (1) expresses a true proposition (i.e. a proposition which is true at the world of the context).

On the basis of (*c*), we might claim that (1) is analytic (i.e. it is true solely in virtue of its meaning). Although as we see from (*b*), (1) rarely or never expresses a necessary proposition. This separation of analyticity and necessity is made possible—even, I hope, plausible—by distinguishing the kinds of entities of which 'is analytic' and 'is necessary' are properly predicated: characters (meanings) are analytic, contents (propositions) are necessary.

The distinction between character and content was unlikely to be noticed before demonstratives came under consideration, because demonstrative-free expressions have a constant character, that is they express the same content in every context. Thus, character becomes an uninteresting complication in the theory.

Though I have spoken above of contexts of utterance, my primary theoretical notion of *content with respect to a context* does not require that the agent of the context utter the expression in question. I believe that there are good reasons for taking this more general notion as fundamental.

I believe that my distinction between character and content can be used to throw light on Kripke's distinction between the *a-priori* and the necessary. Although my distinction lies more purely within logic and semantics, and Kripke's distinction is of a more general epistemic metaphysical character,[1] both seem to me to be of the same *structure*. (I leave this remark in a rather cryptic state.)

The distinction between content and character and the related analysis of demonstratives have certainly been foreshadowed in the literature (though they are original-with-me, in the sense that I did not consciously extract them from prior sources). But to my knowledge they have not previously been cultivated to meet the standards for logical and semantical theories which currently prevail. In particular, Strawson's distinction between the significance (meaningfulness) of a sentence and the statement (proposition) which is expressed in a given use is clearly related.[2] Strawson recognizes that such sentences as 'The *present* King of France is *now* bald' may express different propositions in different utterances, and he identifies the meaningfulness of the sentence with its potential for expressing a true or false proposition in some possible utterance. Though he does not explicitly discuss *the* meaning of the sentence, it is clear that he would not identify such a meaning with any of the propositions expressed by particular utterances. Unfortunately Strawson seems to regard the fact that sentences containing demonstratives can be used to express different propositions as immunizing such sentences against treatment by 'the logician'.

In order to convince myself that it is possible to carry out a consistent analysis of the semantics of demonstratives along the above lines, I have attempted to carry through the program for a version of first-order predicate logic. The result is the following Logic of Demonstratives.

If my views are correct, the introduction of demonstratives into intensional logics will require more extensive reformulation than was thought to be the case.

THE LOGIC OF DEMONSTRATIVES

The *Language* LD is based on first-order predicate logic with identity and descriptions. We deviate slightly from standard formulations in using two sorts of variables, one sort for positions and a second sort for individuals other than positions (hereafter called simply 'individuals').

Primitive Symbols for Two Sorted Predicate Logic

0. Punctuation: (,)
1. (i) An infinite set of individual variables: $\mathscr{V}_i$
   (ii) An infinite set of position variables: $\mathscr{V}_p$
2. (i) An infinite number of *m*-*n*-place predicates, for all natural numbers *m*, *n*
   (ii) The 1-0-place predicate: Exist
   (iii) The 1-1-place predicate: Located
3. (i) An infinite number of *m*-*n*-place *i*-functors (functors which form terms denoting individuals)
   (ii) An infinite number of *m*-*n*-place *p*-functors (functors which form terms denoting positions)
4. Sentential Connectives: $\wedge$, $\vee$, $\neg$, $\rightarrow$, $\leftrightarrow$
5. Quantifiers: $\forall$, $\exists$
6. Definite Description Operator: the
7. Identity: =

Primitive Symbols for Modal and Tense Logic

8. Modal Operators: $\Box$, $\Diamond$
9. Tense Operators: F (it will be the case that)
   P (it has been the case that)
   G (one day ago, it was the case that)

Primitive Symbols for the Logic of Demonstratives

10. Three one place sentential operators:
    N (it is now the case that)
    A (it is actually the case that)
    Y (yesterday, it was the case that)
11. A one place functor: dthat
12. An individual constant (0-0-place *i*-functor): I
13. A position constant (0-0-place *p*-functor): Here

The *well-formed expressions* are of three kinds: formulas, position terms (*p*-terms), and individual terms (*i*-terms).

1. (i) If $\alpha \in \mathscr{V}_i$, then α is an *i*-term.

   (ii) If $\alpha \in \mathscr{V}_p$, then α is a *p*-term.
2. If π is an *m*-*n*-place predicate, $\alpha_1 \ldots \alpha_m$ are *i*-terms, and $\beta_1 \ldots \beta_n$ are *p*-terms, then $\pi\alpha_1 \ldots \alpha_m\beta_1 \ldots \beta_n$ is a formula.
3. (i) If η is an *m*-*n*-place *i*-functor, $\alpha_1 \ldots \alpha_m$, $\beta_1 \ldots \beta_n$ as in 2, then $\eta\alpha_1 \ldots \alpha_m\beta_1 \ldots \beta_n$ is an *i*-term.

   (ii) If η is an *m*-*n*-place *p*-functor, $\alpha_1 \ldots \alpha_m$, $\beta_1 \ldots \beta_n$ as in 2, then $\eta\alpha_1 \ldots \alpha_m\beta_1 \ldots \beta_n$ is a *p*-term.
4. If ϕ, ψ are formulas, then $(\phi \wedge \psi)$, $(\phi \vee \psi)$, $\neg\phi$, $(\phi \rightarrow \psi)$, $(\phi \leftrightarrow \psi)$ are formulas.
5. If ϕ is a formula and $\alpha \in \mathscr{V}_i \cup \mathscr{V}_p$, then $\forall\alpha\phi$, $\exists\alpha\phi$ are formulas.
6. If ϕ is a formula, then

   (i) if $\alpha \in \mathscr{V}_i$, then the α ϕ is an *i*-term.

   (ii) if $\alpha \in \mathscr{V}_p$, then the α ϕ is a *p*-term.

7. If both α, β are either *i*-terms or *p*-terms, then $\alpha = \beta$ is a formula.
8. If ϕ is a formula, then $\Box\phi$, $\Diamond\phi$ are formulas.
9. If ϕ is a formula, then $F\phi$, $P\phi$, $G\phi$ are formulas.
10. If ϕ is a formula, then $N\phi$, $A\phi$, $Y\phi$ are formulas.
11. (i) If α is an *i*-term, then dthat α is an *i*-term.
    (ii) If α is a *p*-term, then dthat α is a *p*-term.

Semantics for LD

DEFINITION. $\mathfrak{A}$ *is an LD Structure* iff there are $\mathscr{C}\mathscr{W}\mathscr{U}\mathscr{P}\mathscr{T}\mathscr{I}$ such that

1. $\mathfrak{A} = \langle \mathscr{C}\mathscr{W}\mathscr{U}\mathscr{P}\mathscr{T}\mathscr{I} \rangle$.
2. $\mathscr{C}$ is a non-empty set (the set of *contexts*, see 10 below).
3. If $c \in \mathscr{C}$, then (i) $c_A \in \mathscr{U}$ (the *agent* of c).
   (ii) $c_T \in \mathscr{T}$ (the *time* of c).
   (iii) $c_P \in \mathscr{P}$ (the *position* of c).
   (iv) $c_w \in \mathscr{W}$ (the *world* of c).
4. $\mathscr{W}$ is a non-empty set (the set of *worlds*).
5. $\mathscr{U}$ is a non-empty set (the set of all *individuals*, see 9 below).
6. $\mathscr{P}$ is a non-empty set (the set of *positions*; common to all worlds).
7. $\mathscr{T}$ is the set of integers (thought of as the *times*; common to all worlds).
8. $\mathscr{I}$ is a function which assigns to each predicate and functor an appropriate *intension* as follows:
   (i) if π is an *m*-*n*-place predicate, $\mathscr{I}_\pi$ is a function such that for each $t \in \mathscr{T}$ and $w \in \mathscr{W}$, $\mathscr{I}_\pi(tw) \subseteq (\mathscr{U}^m \times \mathscr{P}^n)$.
   (ii) If η is an *m*-*n*-place *i*-functor, $\mathscr{I}_\eta$ is a function such that that for each $t \in \mathscr{T}$ and $w \in \mathscr{W}$,

$$\mathscr{I}_\eta(tw) \in (\mathscr{U} \cup \{\dagger\})^{(\mathscr{U}^m \times \mathscr{P}^n)}.$$

   (*Note*: † is a completely alien entity, in neither $\mathscr{U}$ nor

$\mathscr{P}$, which represents an 'undefined' value of the function. In a normal set theory we can take † to be $\{\mathscr{U}, \mathscr{P}\}$.)

(iii) If η is an m-n-place p-functor, $\mathscr{I}_\eta$ is a function such that for each $t \in \mathscr{T}$ and $w \in \mathscr{W}$, $\mathscr{I}_\eta(tw) \in (\mathscr{P} \cup \{\dagger\})^{(\mathscr{U}^m \times \mathscr{P}^n)}$.

9. $i \in \mathscr{U}$ iff $\exists t \in \mathscr{T} \exists w \in \mathscr{W} \langle i \rangle \in \mathscr{I}_{\text{Exists}}(tw)$.
10. If $c \in \mathscr{C}$, then $\langle c_A c_P \rangle \in \mathscr{I}_{\text{Located}}(c_T\ c_W)$.
11. If $\langle i\ p \rangle \in \mathscr{I}_{\text{Located}}(tw)$, then $\langle i \rangle \in \mathscr{I}_{\text{Exists}}(tw)$.

Truth and Denotation in a Context

We write: $\models^{\mathfrak{A}}_{cftw} \phi$ for ϕ when taken in the context c (under the assignment f and in the structure $\mathfrak{A}$) *is true with respect to* the time t and the world w.

We write: $|\alpha|^{\mathfrak{A}}_{cftw}$ for *The denotation of* α when taken in the context c (under the assignment f and in the structure $\mathfrak{A}$) *with respect to* the time t and the world w.

In general we will omit the superscript '$\mathfrak{A}$', and we will assume that the structure $\mathfrak{A}$ is $\langle \mathscr{C} \mathscr{W} \mathscr{U} \mathscr{P} \mathscr{T} \mathscr{I} \rangle$.

DEFINITION. *f is an assignment* (with respect to $\langle \mathscr{C} \mathscr{W} \mathscr{U} \mathscr{P} \mathscr{T} \mathscr{I} \rangle$) iff

$$\exists f_1 f_2 (f_1 \in \mathscr{U}^{\mathscr{V}_1} \ \&\ f_2 \in \mathscr{P}^{\mathscr{V}_p} \ \&\ f = f_1 \cup f_2).$$

DEFINITION. $f^\alpha_x = (f \sim \{\langle \alpha f(\alpha) \rangle\}) \cup \{\langle \alpha x \rangle\}$ (i.e. the assignment which is just like f except that it assigns x to α).

For the following recursive definitions, assume that $c \in \mathscr{C}$, f is an assignment, $t \in \mathscr{T}$, and $w \in \mathscr{W}$.

1. If α is a variable, $|\alpha|_{cftw} = f(\alpha)$.
2. $\models_{cftw} \pi\alpha_1 \ldots \alpha_m \beta_1 \ldots \beta_n$ iff $\langle |\alpha_1|_{cftw} \ldots |\beta_n|_{cftw} \rangle \in \mathscr{I}_\pi(tw)$.

3. If η is neither I nor Here (see 12, 13 below), then
$| \eta\alpha_1 \ldots \alpha_m\beta_1 \ldots \beta_n |_{cftw}$
$$= \begin{cases} \mathscr{I}_\eta(tw)(\langle | \alpha_1 |_{cftw} \ldots | \beta_n |_{cftw}\rangle), \text{ if none of } | \alpha_j |_{cftw} | \beta_k |_{cftw} \text{ are } \dagger \\ \dagger, \text{ otherwise.} \end{cases}$$

4. (i) $\models_{cftw} (\phi \wedge \psi)$ iff $\models_{cftw} \phi \ \& \models_{cftw} \psi$.
(ii) $\models_{cftw} \neg \phi$ iff $\sim \models_{cftw} \phi$.
etc.

5. (i) If $\alpha \in \mathscr{V}_i$, then $\models_{cftw} \forall\alpha\phi$ iff $\forall_i \in \mathscr{U} \models_{cf_i^\alpha tw} \phi$.
(ii) If $\alpha \in \mathscr{V}_p$, then $\models_{cftw} \forall\alpha\phi$ iff $\forall_p \in \mathscr{P} \models_{cf_p^\alpha tw} \phi$.
Similarly for $\exists\alpha\phi$.

6. (i) If $\alpha \in \mathscr{V}_i$, then $|$ the α $\phi |_{cftw}$
$$= \begin{cases} \text{the unique } i \in \mathscr{U} \text{ such that } \models_{cf_i^\alpha tw} \phi, \text{ if there is such.} \\ \dagger, \text{ otherwise.} \end{cases}$$
(ii) Similarly for $\alpha \in \mathscr{V}_p$.

7. $\models_{cftw} \alpha = \beta$ iff $| \alpha |_{cftw} = | \beta |_{cftw}$.

8. (i) $\models_{cftw} \Box \phi$ iff $\forall w' \in \mathscr{W} \models_{cftw'} \phi$.
(ii) $\models_{cftw} \Diamond \phi$ iff $\exists w' \in \mathscr{W} \models_{cftw'} \phi$.

9. (ii) $\models_{cftw} F\phi$ iff $\exists t' \in \mathscr{T}$ such that $t' > t$ and $\models_{cft'w} \phi$.
(ii) $\models_{cftw} P\phi$ iff $\exists t' \in \mathscr{T}$ such that $t' < t$ and $\models_{cft'w} \phi$.
(iii) $\models_{cftw} G\phi$ iff $\models_{cf(t-1)w} \phi$.

10. (i) $\models_{cftw} N\phi$ iff $\models_{cfc_Tw} \phi$.
(ii) $\models_{cftw} A\phi$ iff $\models_{cfc_W} \phi$.
(iii) $\models_{cftw} Y\phi$ iff $\models_{cf(c_T-1)w} \phi$.

11. $|$ dthat $\alpha |_{cftw} = | \alpha |_{cfc_Tc_W}$.

12. $| \mathrm{I} |_{cftw} = c_A$.

13. $| \mathrm{Here} |_{cftw} = c_P$.

Remark 1. Expressions containing demonstratives will, in general, express different concepts in different contexts. We call the concept expressed in a given context the *Content* of the expression in that context. The Content of a sentence in a context is, roughly, the proposition the sentence would express if uttered in that context. This description is not quite accurate on two counts. First, it is important to distinguish an *utterance* from a *sentence-in-a-context*. The former notion is from the theory of speech acts, the latter from semantics. Utterances take time, and utterances of distinct sentences cannot be simultaneous (i.e. in the same context). But in order to develop a logic of demonstratives it seems most natural to be able to evaluate several premisses and a conclusion all in the same context. Thus, the notion of ϕ being true in c and $\mathfrak{A}$ does not require an utterance of ϕ. In particular, c_A need not be uttering ϕ in c_W at c_T. Second, the truth of a proposition is not usually thought of as dependent on a time as well as a possible world. The time is thought of as fixed by the context. If ϕ is a sentence, the more usual notion of the proposition expressed by ϕ-in-c is what is here called the Content of $N\phi$ in c.

Where Γ is either a term or a formula, we write: $\{\Gamma\}^{\mathfrak{A}}_{cf}$ for the Content of Γ in the context c (under the assignment f and in the structure $\mathfrak{A}$).

DEFINITION. (i) If ϕ is a formula, $\{\phi\}_{cf}$ = that function which assigns to each $t \in \mathscr{T}$ and $w \in \mathscr{W}$, Truth if $\models^{\mathfrak{A}}_{cftw} \phi$, and Falsehood otherwise.

(ii) If α is a term, $\{\alpha\}_{cf}$ = that function which assigns to each $t \in \mathscr{T}$ and $w \in \mathscr{W}$, $|\alpha|^{\mathfrak{A}}_{cftw}$

Remark 2. $\models^{\mathfrak{A}}_{cftw} \phi$ iff $\{\phi\}^{\mathfrak{A}}_{cf}(tw)$ = Truth. Roughly speaking, the sentence ϕ taken in the context c is *true with respect to t* and w iff the proposition expressed by ϕ-in-the-context-c would be true at the time t if w were the actual world. In the formal development of pages 74 and 75 it was smoother to ignore the conceptual break marked by the notion of *Content in a context* and to directly define *truth in a context with respect to a possible time and world.* The important conceptual role of the notion of Content is partially indicated by the following two definitions.

DEFINITION. ϕ *is true in the context c* (in the structure $\mathfrak{A}$) iff for every assignment f, $\{\phi\}^{\mathfrak{A}}_{cf}(c_T, c_W)$ = Truth.

DEFINITION. ϕ is valid in LD ($\models \phi$) iff for every LD structure $\mathfrak{A}$, and every context c of $\mathfrak{A}$, ϕ is true in c (in $\mathfrak{A}$).

Remark 3. $\models(\alpha = \text{dthat } \alpha)$, $\models$ N (Located I, Here), $\models$ Exist I, $\sim\models\Box$ $(\alpha = \text{dthat } \alpha)$, $\sim\models\Box$N(Located I, Here), $\sim\models\Box$ (Exist I). In the converse direction we have the usual results in view of the fact that $\models(\Box\phi \rightarrow \phi)$.

DEFINITION. If $\alpha_1 \ldots \alpha_n$ are all the free variables of ϕ in alphabetical order, then *the closure of* ϕ = AN$\forall\alpha_1 \ldots \alpha_n\phi$.

DEFINITION. ϕ *is closed* iff ϕ is equivalent to its closure (in the sense of *Remark* 12, below).

Remark 4. If ϕ is closed, then ϕ is true in c (and $\mathfrak{A}$) iff for every assignment f, time t, and world $w \models^{\mathfrak{A}}_{cftw} \phi$.

DEFINITION. Where Γ is either a term or a formula, *the Content of* Γ *in the context c (in the structure* $\mathfrak{A}$*) is stable* iff for every assignment f, $\{\Gamma\}^{\mathfrak{A}}_{cf}$ is a constant function. (i.e. $\{\Gamma\}^{\mathfrak{A}}_{cf}(tw) = \{\Gamma\}^{\mathfrak{A}}_{cf}(t'w')$, for all t, t', w, w' in $\mathfrak{A}$).

Remark 5. Where ϕ is a formula, α is a term, and β is a variable each of the following has a stable Content in every context (in every structure): *AN*ϕ, dthat α, β, I, Here.

If we were to extend the notion of Content to apply to operators, we would see that all demonstratives have a stable Content in every context. The same is true of the familiar logical constants, although it does not hold for the modal and tense operators (not, at least, according to the foregoing development)

Remark 6. That aspect of the meaning of an expression which determines what its Content will be in each context, we call the *Character* of the expression. Although a lack of knowledge about the context (or perhaps about the structure) may cause one to mistake the Content of a given utterance, the Character of each well formed expression is determined by rules of the language (such as 1–13, pages 74–75 above) which are presumably known to all competent speakers. Our notation '$\{\phi\}^{\mathfrak{A}}_{cf}$' for the Content of an expression gives a natural notation for the Character of an expression, namely '$\{\phi\}$'.

DEFINITION. Where Γ is either a term or a formula the

Character of Γ is that function which assigns to each structure $\mathfrak{A}$, assignment f, and context c of $\mathfrak{A}$, $\{\Gamma\}^{\mathfrak{A}}_{cf}$.

DEFINITION. Where Γ is either a term or a formula, *the Character of Γ is stable* iff for every structure $\mathfrak{A}$, and assignment f the Character of Γ (under f in $\mathfrak{A}$) is a constant function. (i.e. $\{\Gamma\}^{\mathfrak{A}}_{cf} = \{\Gamma\}^{\mathfrak{A}}_{c'f}$ for all c, c' in $\mathfrak{A}$).

Remark 7. A formula or term has a stable Character iff it has the same Content in every context (for each $\mathfrak{A}$, f).

Remark 8. A formula or term has a stable Character iff it contains no essential occurrence of a demonstrative.

Remark 9. The logic of demonstratives determines a sub-logic of those formulas of LD which contain no demonstratives. These formulas (and their equivalents which contain inessential occurrences of demonstratives) are exactly the formulas with a stable Character. The logic of demonstratives brings a new perspective even to formulas such as these. The sub-logic of LD which concerns only formulas of stable Character is not identical with traditional logic. Even for such formulas, the familiar Principle of Necessitation: if $\models \phi$, then $\models \Box\phi$, fails. And so does its tense logic counterpart: if $\models \phi$, then $\models (\neg P \neg \phi \wedge \neg F \neg \phi \wedge \phi)$. From the perspective of LD, validity is truth in every possible *context*. For traditional logic, validity is truth in every possible *circumstance*. Each possible context determines a possible circumstance, but it is not the case that each possible circumstance is part of a possible context. In particular, the fact that each possible context has an agent implies that any possible circumstance in which no individuals exist will not form a part of any possible context. Within LD, a possible context is represented by $\langle \mathfrak{A}, c \rangle$ and a possible circumstance by $\langle \mathfrak{A}, t, w \rangle$. To any $\langle \mathfrak{A}, c \rangle$, there corresponds $\langle \mathfrak{A}, c_T, c_W \rangle$. But it is not the case that to every $\langle \mathfrak{A}, t, w \rangle$ there exists a context c of $\mathfrak{A}$ such that $t = c_T$ and $w = c_W$. The result is that in LD such sentences as $\exists x$ Exist x and $\exists x \exists p$ Located x, p are valid, although they would not be so regarded in traditional logic. At least not in the neo-traditional logic that countenances empty worlds. Using the semantical developments of pages 73–75, we can define this traditional sense of validity (for formulas which do not contain demonstratives) as follows. First note that by *Remark* 7, if ϕ has a stable Character

$\models^{\mathfrak{A}}_{cftw} \phi$ iff $\models^{\mathfrak{A}}_{c'ftw} \phi$.

Thus for such formulas we can define,

ϕ *is true at tw (in $\mathfrak{A}$)* iff for every assignment f and every context c

$$\models^{\mathfrak{A}}_{cftw} \phi.$$

The neo-traditional sense of validity is now definable as follows:

$\models_T \phi$ iff for all structures $\mathfrak{A}$, times t, and worlds w, ϕ is true at tw (in $\mathfrak{A}$).

(Properly speaking, what I have called the neo-traditional sense of validity is the notion of validity now common for a quantified S5 modal tense logic with individual variables ranging over possible individuals and a predicate of existence.) Adding the subscript 'LD' for explicitness, we can now state some results.

(i) If ϕ contains no demonstratives, if $\models_T \phi$, then $\models_{LD} \phi$.

(ii) $\models_{LD} \exists x$ Exist x, but $\sim \models_T \exists x$ Exist x.

Of course $\Box\exists x$ Exist x is not valid even in LD. Nor are its counterparts, $\neg F \neg \exists x$ Exist x and $\neg P \neg \exists x$ Exist x.

This suggests that we can transcend the context-oriented perspective of LD by generalizing over times and worlds so as to capture those possible circumstances $\langle\mathfrak{A}, t, w\rangle$ which do not correspond to any possible contexts $\langle\mathfrak{A}, c\rangle$. We have the following result:

(iii) If ϕ contains no demonstratives
$\models_T \phi$ iff $\models_{LD} \Box(\neg P \neg \phi \wedge \neg F \neg \phi \wedge \phi)$.

Although our definition of the neo-traditional sense of validity was motivated by consideration of demonstrative-free formulas, we could apply it also to formulas containing essential occurrences of demonstratives. To do so would nullify the most interesting features of the logic of demonstratives. But it raises the question: can we express our new sense of validity in terms of the neo-traditional sense? This can be done:

(iv) $\models_{LD} \phi$ iff $\models_T \text{AN}\phi$.

Remark 10. Rigid designators (in the sense of Kripke) are terms with a stable Content. Since Kripke does not discuss demonstratives, his examples all have, in addition, a stable Character (by *Remark 8*). Kripke claims that for proper names α, β it may happen that $\alpha = \beta$, though not *a-priori*, is nevertheless necessary. This, in spite of the fact that the names α, β may be introduced by means of descriptions α', β' for which $\alpha' = \beta'$ is not necessary. An analogous situation holds in LD. Let α', β' be definite descriptions (without free variables) such that $\alpha' = \beta'$ is not *a-priori*, and consider the rigid terms dthat α' and dthat β' which are formed from them. We know that $\models$(dthat α' = dthatβ' $\leftrightarrow$ $\alpha' = \beta'$). Thus, if $\alpha' = \beta'$ is not *a-priori*, neither is dthatα' = dthat β'. But, since $\models$[dthat α' = dthatβ' $\rightarrow$ $\Box$(dthatα' = dthatβ')], it may happen that dthatα' = dthatβ' is necessary. The converse situation can also be illustrated in LD. Since (α' = dthatα') is valid (see *Remark 3*), it is surely capable of being known *a-priori*. But if α' lacks a stable Content (in some context c), $\Box$ (α' = dthatα') will be false.

Remark 11. Our *o-o*-place *i*-functors are not proper names, in the sense of Kripke, since they do not have a stable Content. But they can easily be converted by means of the stabilizing influence of dthat. Even dthat α lacks a stable Character. The process by which such expressions are converted into expressions with a stable Character is 'dubbing'—a form of definition in which context may play an essential role. The means to deal with such context indexed definitions is not available in our object language.

There would, of course, be no difficulty in supplementing our language with a syntactically distinctive set of *o-o*-place *i*-functors whose semantics requires them to have both a stable Character and a stable Content in every context. Variables already behave this way; what is wanted is a class of constants that behave, in these respects, like variables.

The difficulty comes in expressing the definition. My thought is that when a name, like 'Bozo', is introduced by someone saying, in some context c^*, 'Let's call the Governor, "Bozo"', we have a context indexed definition of the form: $A =_{c^*} \alpha$, where A is a new constant (here, 'Bozo') and α is some term whose denotation depends on context (here, 'the Governor'). The intention of such a dubbing is, presumably, to induce the semantical clause: for all c, $\{A\}^{\mathfrak{A}}_{cf} = \{\alpha\}^{\mathfrak{A}}_{c^*f}$. Such a clause gives A a stable Character. The context indexing is required by the fact that the Content of α (the 'definiens') may vary from context to context. Thus the same semantical clause is not induced by taking either $A = \alpha$ or even $A = \text{dthat } \alpha$ as an axiom.

I think it likely that such definitions play a practically (and perhaps theoretically) indispensable role in the growth of language, allowing us to introduce a vast stock of names on the basis of a meagre stock of demonstratives and some ingenuity in the staging of demonstrations.

Perhaps such introductions should not be called 'definitions' at all, since they essentially enrich the expressive power of the language. What a nameless man may express by 'I am hungry' may be inexpressible in remote contexts. But once he says 'Let's call me "Bozo" ' his Content is accessible to us all.

Remark 12. The strongest form of logical equivalence between two formulas ϕ and ϕ' is sameness of Character, $\{\phi\} = \{\phi'\}$. This form of synonymy is expressible in terms of validity:

$$\{\phi\} = \{\phi'\} \text{ iff } \models \Box[\neg P \neg(\phi \leftrightarrow \phi') \wedge \neg F \neg(\phi \leftrightarrow \phi') \wedge (\phi \leftrightarrow \phi')].$$

[Using *Remark* 9 (iii) and dropping the condition, which was stated only to express the intended range of applicability of $\models_T$, we have: $\{\phi\} = \{\phi'\}$ iff $\models_T(\phi \leftrightarrow \phi')$.] Since definitions of the usual kind (as opposed to dubbings) are intended to introduce a short expression as a mere abbreviation of a longer one, the Character of the defined sign should be the same as the Character of the definiens. Thus, with LD, definitional axioms must take the form indicated above.

Remark 13. If β is a variable of the same sort as the term α but is not free in α, then $\{\text{dthat } \alpha\} = \{\text{the } \beta\ AN(\beta = \alpha)\}$. Thus for every formula ϕ, there can be constructed a formula ϕ' such that ϕ' contains no occurrence of dthat and $\{\phi\} = \{\phi'\}$.

Remark 14. Y (yesterday) and G (one day ago) superficially resemble one another in view of the fact that $\models (Y\phi \leftrightarrow G\phi)$. But the former is a demonstrative whereas the latter is an iterative temporal operator. 'One day ago it was the case that one day ago it was the case that John yawned' means that John yawned the day before yesterday. But 'Yesterday it was the case that yesterday it was the case that John yawned' is only a stutter.

POSSIBLE REFINEMENTS

(1) The primitive predicates and functors of first-order predicate

logic are all taken to be extensional. Alternatives are possible.

(2) Many conditions might be added on $\mathscr{P}$; many alternatives might be chosen for $\mathscr{T}$. If the elements of $\mathscr{T}$ do not have a natural relation to play the role of $<$, such a relation must be added to the structure.

(3) When K is a set of LD formulas, $K \models \phi$ is easily defined in any of the usual ways.

(4) Aspects of the contexts other than c_A, c_P, c_T, and c_W would be used if new demonstratives (e.g. pointings, 'You', etc.) were added to the language. (Note that the subscripts A, P, T, W are external parameters. They may be thought of as functions applying to contexts, with c_A being the value of A for the context c.)

(5) Special continuity conditions through time might be added for the predicate Exists.

(6) If individuals lacking positions are admitted as agents of contexts, 3(iii) of page 73 should be weakened to $c_P \in \mathscr{P} \cup \{\dagger\}$. It would no longer be the case that $\models$ Located I, Here. If individuals also lacking temporal location (disembodied minds?) are admitted as agents of contexts, a similar weakening is required of 3(ii). In any case it would still be true that $\models$ Exist I.

Notes

* This paper was originally composed in two parts. The formal 'Logic of Demonstratives' was first presented at the Irvine Summer Institute on the Philosophy of Language in 1971. It was expanded in 1973. The initial discursive material was written on 20 April 1973, as part of a research proposal. This paper was intended as a companion piece to and progress report on the material in 'Dthat'. (D. Kaplan, 'Dthat' in Peter Cole (ed.), *Syntax and Semantics*, 9, *Pragmatics* (Academic Press: New York, 1978), 221–43. Also reprinted in P. French *et al.* (eds), *Contemporary Perspectives in the Philosophy of Language* (Minneapolis: University of Minnesota Press, 1979), pp. 383–400.) A more extensive presentation occurs in my manuscript 'Demonstratives', in J. Almog. J. Perry, and H. Wettstein (eds.), *Themes from Kaplan* (Oxford University Press, forthcoming). This work was supported by the National Science Foundation.

1. S. Kripke, 'Naming and Necessity', in Donald Davidson and Gilbert Harman (eds.), *Semantics of Natural Language* (Dordrecht: Reidel, 1972) 253–355; Addenda, pp. 763–9.
2. P. Strawson, *Introduction to Logical Theory*, (New York: *John Wiley & Sons*, 1952).

VI

THE PROBLEM OF THE ESSENTIAL INDEXICAL

JOHN PERRY

I ONCE followed a trail of sugar on a supermarket floor, pushing my trolly down the aisle on one side of a tall counter and back along the aisle on the other, seeking the shopper with the torn bag to tell him he was making a mess. With each trip around the counter, the trail became thicker. But I seemed uanable to catch up. Finally it dawned on me. I was the shopper I was trying to catch.

I believed at the outset that the shopper with a torn bag was making a mess. And I was right. But I did not believe that I was making a mess. That seems to be something I came to believe. And when I came to believe that, I stopped following the trail around the counter, and rearranged the torn bag in my trolley. My change in beliefs seems to explain my change in behaviour. My aim in this paper is to make a key point about the characterization of this change, and of beliefs in general.

At first characterizing the change seems easy. My beliefs changed, did they not in that I came to have a new one, namely, *that I am making a mess*? But things are not so simple.

The reason they are not is the importance of the word "I" in my expression of what I came to believe. When we replace it with other designations of me, we no longer have an explanation of my behaviour and so, it seems, no longer an attribution of the same belief. It seems to be an *essential* indexical. But without such a replacement, all we have to identify the belief is the sentence "I am making a mess". But that sentence by itself does not seem to identify the crucial belief, for if someone else had said it, they would have expressed a different belief, a false one.

John Perry, 'The Problem of the Essential Indexical', *Noûs* 13 (1979), 3–21.

I argue that the essential indexical poses a problem for various otherwise plausible accounts of belief. I first argue that it is a problem for the view that belief is a relation between subjects and propositions conceived as bearers of truth and falsity. The problem is not solved merely by replacing or supplementing this with a notion of *de re* belief. Nor is it solved by moving to a notion of a proposition which, rather than true or false absolutely, is only true or false at an index or in a context (at a time, for a speaker, say). Its solution requires us to make a sharp distinction between objects of belief and belief states, and to realize that the connection between them is not so intimate as might have been supposed.[1]

LOCATING BELIEFS

I want to introduce two more examples. In the first a professor, who desires to attend the department meeting on time, and believes correctly that it begins at noon, sits motionless in his office at that time. Suddenly he begins to move. What explains his action? A change in belief. He believed all along that the department meeting starts at noon; he came to believe, as he would have put it, that it starts *now*.

The author of the book *Hiker's Guide to the Desolation Wilderness* stands in the wilderness beside Gilmore Lake, looking at the Mt. Tallac trail as it leaves the lake and climbs the mountain. He desires to leave the wilderness. He believes that the best way out from Gilmore Lake is to follow the Mt. Tallac trail up the mountain to Cathedral Peaks trail, on to the Floating Island trail, emerging at Spring Creek Tract Road. But he does not move. He is lost. He is not sure whether he is standing beside Gilmore Lake, looking at Mt. Tallac, or beside Clyde Lake, looking at Jack's peak, or beside Eagle Lake, looking at one of the Maggie peaks. Then he begins to move along the Mt. Tallack trail. If asked, he would have explained the crucial change in his beliefs this way: "I came to believe that *this* is the Mt. Tallac trail and *that* is Gilmore Lake."

In these three cases the subjects, in explaining their actions, would use indexicals to characterize certain beliefs they came to have. These indexicals are essential, in that replacement of them by other terms destroys the force of the explanation, or at least requires certain assumptions to be made to preserve it.

Suppose I had said, in the manner of de Gaulle, "I came to believe that John Perry is making a mess." I would no longer have explained why I stopped and looked in my own trolley. To explain that I would have to add, "and I believe that I am John Perry", bringing in the indexical again. After all, suppose I had really given my explanation in the manner of de Gaulle, and said "I came to believe that de Gaulle is making a mess." That would not have explained my stopping at all. But it really would have explained it every bit as much as "I came to believe John Perry is making a mess." For if I added "and I believe that I am de Gaulle" the explanations would be on par. The only reason "I came to believe John Perry is making a mess" seems to explain my action is our natural assumption that I did believe I was John Perry and did not believe I was de Gaulle. So replacing the indexical 'I' with another term designating the same person really does, as claimed, destroy the explanation.

Similarly, our professor, as he set off down the hall, might say "I believe the meeting starts at noon", rather than "I believe the meeting starts now". In accepting the former as an explanation, we would be assuming he believes it is *now* noon. If he believed it was now 5 p.m., he would not have explained his departure by citing his belief that the meeting starts at noon, unless he was a member of a department with very long meetings. After all, he believed that the meeting started at noon all along, so that belief can hardly explain a change in his behaviour. Basically similar remarks apply to the lost author.

I shall use the term "locating beliefs" to refer to one's beliefs about where one is, when it is, and who one is. Such beliefs seem essentially indexical. Imagine two lost campers who trust the same guidebook but disagree about where they are. If we were to try to characterize the beliefs of these campers without the use of indexicals, it would seem impossible to bring out this disagreement. If, for example, we characterized their beliefs by the set of "eternal sentences" drawn from the guidebook they would mark "true", there is no reason to suppose that the sets would differ. They could mark all of the same sentences "true", and still disagree in their locating beliefs. It seems that there has to be some indexical element in the characterization of their beliefs to bring out this disagreement. But as we shall see there is no room for this indexical element in the traditional way of looking at belief, and even when its necessity is recognized, it is not easy to see how to fit it in.

THE DOCTRINE OF PROPOSITIONS

I shall first consider how the problem appears to a traditional way of thinking of belief. The doctrines I describe were held by Frege, but I shall put them in a way that does not incorporate his terminology or the details of his view.[2] This traditional way, which I call the "doctrine of propositions", has three main tenets. The first is that belief is a relation between a subject and an object, the latter being denoted, in a canonical belief report, by a that-clause. So "Carter believes that Atlanta is the capital of Georgia" reports that a certain relation, *believing*, obtains between Carter and a certain object—at least in a suitably wide sense of object—*that Atlanta is the capital of Georgia*. These objects are called *propositions*.

The second and the third tenets concern such objects. The second is that they have a truth-value in an absolute sense, as opposed to being true for a person or at a time. The third has to do with how we individuate them. It is necessary, for *that S* and *that S′* to be the same, that they have the same truth-value. But it is not sufficient, for *that the sea is salty* and *that milk is white* are not the same proposition. It is necessary that they have the same truth condition, in the sense that they attribute to the same objects the same relation. But this also is not sufficient, for *that Atlanta is the capital of Georgia* and *that Atlanta is the capital of the largest state east of the Mississippi* are not the same proposition. Carter, it seems, might believe the first but not the second. Propositions must not only have the same truth-value, and concern the same objects and relations, but also involve the same concepts. For Frege, this meant that if *that S* = *that S′*, *S* and *S′* must have the same sense. Others might eschew senses in favour of properties and relations, others take concepts to be just words, so that sameness of propositions is just sameness of sentences. What these approaches have in common is the insistence that propositions must be individuated in a more "fine-grained" way than is provided by truth-value or the notion of truth conditions employed above.

THE PROBLEM

It is clear that the essential indexical is a problem for the doctrine of propositions. What answer can it give the question, "What did I come to believe when I straightened up the sugar?" The sentence "I am making a mess" does not identify a proposition. For this sentence is not true or false absolutely, but only as said by one person or another; had another shopper said it when I did, he would have been wrong. So the sentence by which I identify what I came to believe does not identify, by itself, a proposition. There is a *missing conceptual ingredient*: a sense for which I am the reference, or a complex of properties I alone have, or a singular term that refers to no one but me. To identify the proposition I came to believe, the advocate of the doctrine of propositions must identify this missing conceptual ingredient.

An advocate of the doctrine of propositions, his attention drawn to indexicals, might take this attitude towards them: they are communicative shortcuts. Just before I straightened up the bag I must have come to believe some propositions with the structure α *is making a mess*, where α is some concept which I alone "fit" (to pick a phrase neutral among the different notions of a concept). When I say "I believe I am making a mess", my hearers know that I believe some such proposition of this form; which one in particular is not important for the purposes at hand.

If this is correct, we should be able to identify the proposition I came to believe, even if doing so is not necessary for ordinary communicative purposes. But then the doctrine of propositions is in trouble, for any candidate will fall prey to the problems mentioned above. If *that* α *is making a mess* is what I came to believe, then "I came to believe that *A* is making a mess", where *A* expressed α, should be an even better explanation than the original, where I used "I" as a communicative shortcut. But, as we saw, any such explanation will be defective, working only on the assumption that I believed that I was α.

To this it might be replied that though there may be no replacement for "I" that generally preserves explanatory force, all that needs to be claimed is that there is such a replacement on each occasion. The picture is this. On each occasion that I use "I", there is some concept I have in mind that fits me uniquely, and which is the missing conceptual ingredient in the proposition that remains incompletely identified when I characterize my beliefs. The concept I use to think of myself is not necessarily the same each time I do so, and of course I must use a different one than others do, since it must fit me and not them. Because there is no general way of replacing the "I" with a term that gets at the missing ingredient, the challenge to do so in response to a particular example is temporarily embarrassing. But the doctrine of propositions does not require a general answer.

This strategy does not work for two reasons. First, even if I was thinking of myself as, say, the only bearded philosopher in a Safeway store west of the Mississippi, the fact that I came to believe that the only such philosopher was making a mess explains my action only on the assumption that I believed that I was the only such philosopher, which brings in the indexical again. Second, in order to provide me with an appropriate proposition as the object of belief, the missing conceptual ingredient will have to fit me. Suppose I was thinking of myself in the way described, but that I was not bearded and was not in a Safeway store—I had forgotten that I had shaved and gone to the A & P instead. Then the proposition supplied by this strategy would be false, while what I came to believe, *that I was making a mess*, was true.

This strategy assumes that whenever I have a belief I would characterize by using a sentence with an indexical *d*,

I believe that . . . d . . .

that there is some conceptual ingredient *c*, such that it is also true that,

I believe that *d* is *c*

and that, on this second point, I am right. But there is no reason to believe this would always be so. Each time I say "I believe it is *now* time to rake the leaves", I need not have some concept that uniquely fits the time at which I speak.

From the point of view of the doctrine of propositions, belief reports such as "I believe that I am making a mess" are deficient, for there is a missing conceptual ingredient. From the point of view of locating beliefs, there is something lacking in the propositions offered by the doctrine, a missing indexical ingredient.

The problem of the essential indexical reveals that something is badly wrong with the traditional doctrine of propositions. But the traditional doctrine has its competitors anyway, in response to philosophical pressures from other directions. Perhaps attention to these alternative or supplementary models of belief will provide a solution to our problem.

DE RE BELIEF

One development in the philosophy of belief seems quite promising in this respect. It involves qualifying the third tenet of the doctrine of propositions, to allow a sort of proposition individuated by an object or sequence of objects, and a part of a proposition of the earlier sort. The motivation for this qualification or supplementation comes from belief reports which give rise to the same problem, that of the missing conceptual ingredient, as does the problem of the essential indexical.

The third tenet of the doctrine of propositions is motivated by the failure of substitutivity of co-referential terms within the that-clause following "believes". But there seems to be a sort of belief report, or a way of understanding some belief reports, that allows such substitution, and such successful substitution becomes a problem for a theory designed to explain its failure. For suppose Patrick believes that, as he would put it, the dean is wise. Patrick does not know Frank, much less know that he lives next to the dean, and yet I might in certain circumstances say "Patrick believes Frank's neighbour is wise." Or I might say "There is someone whom Patrick believes to be wise', and later on identify that someone as "Frank's neighbour". The legitimacy of this cannot be understood on the unqualified doctrine of propositions; I seem to have gone from one proposition, *that the dean of the school is wise*, to another, *that Frank's neighbour is wise*; but the fact that Patrick believes the first seems to be no reason he should believe the second. And the quantification into the belief report seems to make no sense at all on the doctrine of propositions, for the report does not relate Patrick to an individual known variously as "the dean" and "Frank's neighbour", but only with a concept expressed by the first of these terms.

The problem here is just that of a missing conceptual ingredient. It looked in the original report as if Patrick was being said to stand in the relation of belief to a certain proposition, a part of which was a conceptual ingredient expressed by the words "the dean". But if I am permitted to exchange those words for others, "Frank's neighbour", which are not conceptually equivalent, then apparently the initial part of the proposition he was credited with belief in was not the conceptual ingredient identified by "the dean" after all. So what proposition was it Patrick was originally credited with belief in? And "There is someone such that Patrick believes that he is wise" seems to credit Patrick with belief in a proposition, without telling us which one. For after the "believes" we have only "he is wise", where the "he" does not give us an appropriate conceptual ingredient, but functions as a variable ranging over individuals.

We do seem in some circumstances to allow such substitutivity, and make ready sense of quantification into belief reports. So the doctrine of propositions must be qualified. We can look upon this sort of belief as involving a relation to a new sort of proposition, consisting of an object or sequence of objects and a conceptual ingredient, a part of a proposition of the original kind, or what we might call an "open proposition". This sort of belief and this kind of proposition we call "*de re*", the sort of belief and the sort of proposition that fits the original doctrine, "*de dicto*". Taken this way we analyse "Patrick believes that the dean of the school is wise", as reporting a relation between Patrick and a proposition consisting of a certain person variously describable as "the dean" and "Frank's neighbour" and something, *that x is wise*, which would yield a proposition with the addition of an appropriate conceptual ingredient. Since the dean himself, and not just a concept expressed by the words "the dean," is involved, substitution holds and quantification makes sense.

Here, as in the case of the essential indexical, we were faced with a missing conceptual ingredient. Perhaps, then, this modification of the third tenet will solve the earlier problem as well. But it will not. Even if we suppose—as I think we should—that when I said "I believe that I am making a mess" I was reporting a *de re* belief, our problem will remain.

One problem emerges when we look at accounts that have been offered of the conditions under which a person has a *de re* belief. The most influential treatments of *de re* belief have tried to explain it in terms of *de dicto* belief or something like it. Some terminological regimentation is helpful here. Let us couch reports of *de re* belief in the terms "*X* believes of *a* that he is so and so", reserving the simpler "*X* believes that *a* is so-and-so" for *de dicto* belief. The simplest account of *de re* belief in terms of de dicto belief is this:

> *X* believes of *y* that he is so-and-so

just in case

> There is a concept α such that α fits *y* and *X* believes that α is so-and-so.

Now it is clear that if this is our analysis of *de re* belief, the problem of the essential indexical is still with us. For we are faced with the same problem we had before. I can believe that I am making a mess, even if there is no concept α such that I alone fit α and I believe that α is making a mess. Since I do not have any *de dicto* belief of the sort, on this account I do not have a *de re* belief of the right sort either. So, even allowing *de re* belief, we still do not have an account of the belief I acquired.

Now this simple account of *de re* belief has not won many adherents, because it is commonly held that *de re* belief is a more interesting notion than it allows. This proposal trivializes it. Suppose Nixon is the next President. Since I believe that the next president will be the next President, I would on this proposal believe of Nixon that he is the next President, even though I am thoroughly convinced that Nixon will not be the next President.[3]

To get a more interesting or useful notion of *de re* belief, philosophers have suggested that there are limitations on the conceptual ingredient involved in the *de dicto* belief which yields the *de re* belief. Kaplan, for example, requires not only that there be some α such that I believe that α will be the next President and that α denotes Nixon, for me to believe of Nixon that he will be the next President, but also that α be a *vivid name of Nixon for me*.[4] Hintikka requires that α denote the same individual in every possible world compatible with what I believe.[5] Each of these philosophers explains these notions in such a way that in the circumstances imagined, I would not believe of Nixon that he is the next President.

However well these proposals deal with other phenomena connected with *de re* belief, they cannot help with the problem of the essential indexical. They tighten the requirements laid down by the original proposal, but those were apparently already too restrictive. If in order to believe that I am making a mess I need not have any conceptual ingredient α that fits me, *a fortiori* I am not required to have one that is a vivid name of myself for me, or one that picks out the same individual in every possible world compatible with what I believe.

Perhaps this simply shows that the approach of explaining *de re* belief in terms of *de dicto* belief is incorrect. I think it does show that. But even so, the problem remains. Suppose we do not insist on an account of *de re* belief in terms of *de dicto* belief, but merely suppose that whenever we ascribe a belief, and cannot find a suitable complete proposition to serve as the object because of a missing conceptual ingredient, we are dealing with *de re* belief. Then we will ascribe a *de re* belief to me in the supermarket—I believed *of* John Perry that he was making a mess. But it will not be my having such a *de re* belief that explains my action.

Suppose there were mirrors at either end of the counter so that as I pushed my trolley down the aisle in pursuit I saw myself in the mirror. I take what I see to be the reflection of the messy shopper going up the aisle on the other side, not realizing that what I am really seeing is a reflection of a reflection of myself. I point and say, truly, “I believe that he is making a mess.” In trying to find a suitable proposition for me to believe, we would be faced with the same sorts of problems we had with my earlier report, in which I used “I” instead of “he”. We would not be able to eliminate an indexical element in the term referring to me. So here we have *de re* belief; I believe of John Perry that he is making a mess. But then that I believe of John Perry that he is making a mess does not explain my stopping; in the imagined circumstances I would accelerate, as would the shopper I was trying to catch. But then, even granting that when I say “I believe that I am making a mess” I attribute to myself a certain *de re* belief, the belief of John Perry that he is making a mess, our problem remains.

If we look at it with the notion of a locating belief in mind, the failure of the introduction of *de re* belief to solve our problems is not surprising. *De re* propositions remain non-indexical. Propositions individuated in part by objects remain as insensitive to what is essential in locating beliefs as those individuated wholly by concepts. Saying that I believed of John Perry that he was making a mess leaves out the crucial change, that I came to think of the messy shopper not merely as the shopper with the torn bag or the man in the mirror, but as *me*.

RELATIVIZED PROPOSITIONS

It seems that to deal with essential indexicality we must somehow incorporate the indexical element in what is believed, the object of belief. If we do so, we come up against the second tenet of the doctrine of propositions, that such objects are true or false absolutely. But the tools for abandoning this tenet have been provided in recent treatments of the semantics of modality, tense, and indexicality. So this seems a promising direction.

In possible worlds semantics for necessity and possibility we have the notion of truth at a world. In a way this does not involve a new notion of a proposition and in a way it does. When Frege insisted that his "thoughts" were true or false absolutely, he did not mean that they had the same truth-value in all possible worlds. Had he used a possible worlds framework, he would have had their truth-values vary from world to world, and simply insisted on a determinate truth-value in each world and in particular in the actual world. In a way then, taking propositions to be functions from possible worlds to truth-values is just a way of looking at the old notion of a proposition.

Still, this way of looking at it invites generalization that takes us away from the old notion. From a technical point of view, the essential idea is that a proposition is or is represented by a function from an index to a truth-value; when we get away from modality, this same technical idea may be useful, though something other than possible worlds are taken as indices. To deal with temporal operators, we can use the notion of truth at a time. Here the indices will be times, and our propositions will be functions from times to truth-values. For example, *that Elizabeth is Queen of England* is a proposition true in 1960 but not in 1940. Hence "At sometime or other Elizabeth is Queen of England" is true, *simpliciter*.[6]

Now consider "I am making a mess". Rather than thinking of this as partially identifying an absolutely true proposition, with the "I" showing the place of the missing conceptual ingredient, why not think of it as *completely identifying* a new-fangled proposition, that is true or false only *at a person*? More precisely, it is one that is true or false at a time and a person, since though true when I said it, it has since occasionally been false.

If we ignore possibility and necessity, it seems that regarding propositions as functions to truth-values from indices which are pairs of persons and times will do the trick, and that so doing will allow us to exploit relations between elements within the indices to formulate rules which bring out differences between indexicals. "I am tired now" is true at the pair consisting of the person *a* and the time *t* if and only if *a* is tired at *t*, while "You will be tired" is true at the same index if and only if the addressee of *a* at *t* is tired at some time later than *t*.

Does this way of looking at the matter solve the problem of the essential indexical? I say "I believe that I am making a mess". On our amended doctrine of propositions, this ascribes a relation between me and *that I am making a mess*, which is a function from indices to truth-values. The belief report seems to completely specify the relativized proposition involved; there is no missing conceptual ingredient. So the problem must be solved.

But it is not. I believed that a certain proposition, *that I am making a mess* was true—true for me. So belief that this proposition was true for me then does not differentiate me from the other shopper, and can not be what explains my stopping and searching my trolley for the torn bag. Once we have adopted these new-fangled propositions, which are only true at times for persons, we have to admit also that we believe them as true for persons at times, and not absolutely. And then our problem returns.

Clearly an important distinction must be made. All believing is done by persons at times, or so we may suppose. But the time of belief and the person doing the believing cannot be generally identified with the person and time relative to which the proposition believed is held true. You now believe that *that I am making a mess* was true for me, then, but you certainly do not believe it is true for you now, unless you are reading this in a supermarket. Let us call *you* and *now* the context of belief, and *me* and *then* the context of evaluation. The context of belief may be the same as the context of evaluation, but need not be.

Now the mere fact that I believed the proposition *that I am making a mess* to be true for someone at some time did not explain my stopping the trolley. You believe so now, and doubtless have no more desire to mess up supermarkets than I did. But you are not bending over to straighten up a bag of sugar.

The fact that I believed this proposition true for Perry at the time he was in the supermarket does not explain my behaviour either. For so did the other shopper. And you also now believe this proposition was true for Perry at the time he was in the supermarket.

The important difference seems to be that for me the context of belief was just the context of evaluation, but for the other shopper it was not and for you it is not. But this does not do the trick either.

Consider our tardy professor. He is doing research on indexicals, and has written on the board "My meeting starts now". He believes that the proposition expressed by this sentence is true at noon for him. He has believed so for hours, and at noon the context of belief comes to be the context of evaluation. These facts give us no reason to expect him to move.

Or suppose I think to myself that the person making the mess should say so. Turning my attention to the proposition, I certainly believe *that I am making a mess* is true for the person who ought to be saying it (or the person in the mirror, or the person at the end of the trail of sugar) at that time. The context of evaluation is just the context of belief. But there is no reason to suppose I would stop my trolley.

One supposes that in these cases the problem is that the context of belief is not believed to be the context of evaluation. But formulating the required belief will simply bring up the problem of the essential indexical again. Clearly and correctly we want the tardy professor, when he finally sees he must be off to the meeting, to be ready to say "I believe that the time at which it is true *that the meeting starts now* is now." On the present proposal, we analyse the belief he thereby ascribes to himself as belief in the proposition *that the time at which it is true that the meeting starts now is now*. But he certainly can believe at noon, that this whole proposition is true at noon, without being ready to say "It's starting now" and leave. We do not yet have a solution to the problem of the essential indexical.

LIMITED ACCESSIBILITY

One may take all that has been said so far as an argument for the existence of a special class of propositions, propositions of limited accessibility. For what have we really shown? All attempts to find a formula of the form "*A* is making a mess", with which any of us at any time could express what I believed, have failed. But one might argue that we can hardly suppose that there was not anything that I believed; surely I believed just that proposition which I expressed, on that occasion, with the words "I am making a mess".That we cannot find a sentence that always expresses this proposition when said by anyone does not show that it does not exist. Rather it should lead us to the conclusion that there is a class of propositions which can only be expressed in special circumstances. In particular, only I could express the proposition I expressed when I said "I am making a mess." Others can see, perhaps by analogy with their own case, that there is a proposition that I express, but it is in a sense inaccessible to them.

Similarly, at noon on the day of the meeting, we could all express the proposition the tardy professor expressed with the words "The meeting starts now". But once that time has past, the proposition becomes inaccessible. We can still identify it as the proposition which was expressed by those words at that time. But we cannot express it with those words any longer, for with each passing moment they express a different proposition. And we can find no other words to express it.

The advocate of such a stock of propositions of limited accessibility may not need to bring in special propositions accessible only at certain places. For it is plausible to suppose that other indexicals can be eliminated in favour of "I" and "now". Perhaps "That is Gilmore Lake" just comes to "What I see now in front of me is Gilmore Lake". But elimination of either "I" or "now" in favour of the other seems impossible.

Such a theory of propositions of limited accessibility seems acceptable, even attractive, to some philosophers.[7] Its acceptability or attractiveness will depend on other parts of one's metaphysics; if one finds plausible reasons elsewhere for believing in a universe that has, in addition to our common world, myriads of private perspectives, the idea of propositions of limited accessibility will fit right in.[8] I have no knockdown argument against such propositions, or the metaphysical schemes that find room for them. But I believe only in a common actual world. And I do not think the phenomenon of essential indexicality forces me to abandon the view.

THE OBVIOUS SOLUTION?

Let us return to the device of the true-false exam. Suppose the lost author had been given such an exam before and after he figured out where he was. Would we expect any differences in his answers? Not so long as the statements contained no indexicals. "Mt Tallac is higher than either of the Maggie Peaks" would have been marked the same way before and after, the same way he would have marked it at home in Berkeley. His mark on that sentence would tell us nothing about where he thought he was. But if the exam were to contain such sentences as "That is Gilmore Lake in front of me", we would expect a dramatic change from "False" or "Unsure" to "True".

Imagine such an exam given to various lost campers in different parts of the Wilderness. We could classify the campers by their answers, and such a classification would be valuable for prediction and explanation. Of all the campers who marked "This is Gilmore Lake" with "True", we would say they believed that they were at Gilmore Lake. And we should expect them to act accordingly; if they possessed the standard guidebook, and wished to leave the Wilderness, we might expect what is, given one way of looking at it, the same behaviour: taking the path up the mountain above the shallow end of the lake before them.

Now consider all the good-hearted people who have ever been in a supermarket, noticed sugar on the floor, and been ready to say "I am making a mess." They all have something important in common, something that leads us to expect their next action to be that of looking into their grocery trolleys in search of the torn bag. Or consider all the responsible professors who have ever uttered "The department meeting is starting now." They too have something important in common; they are in a state which will lead those just down the hall to go to the meeting, those across campus to curse and feel guilty, those on leave to smile.

What the members within these various groups have in common is not what they believe. There is no *de dicto* proposition that all the campers or shoppers or professors believe. And there is no person whom all the shoppers believe to be making a mess, no lake all the campers believe to be Gilmore Lake, and no time at which all the professors believe their meetings to be starting.

We are clearly classifying the shoppers, campers, and professors into groups corresponding to what we have been calling "relativized propositions"—abstract objects corresponding to sentences containing indexicals. But what members of each group have in common, which makes the groups significant, is not belief that a certain relativized proposition is true. Such belief, as we saw, is belief that such a proposition is true at some context of evaluation. Now all of the shoppers believe that *that I am making a mess* is true at some context of evaluation or other, but so does everyone else who has ever given it a moment's thought. And similar remarks apply to the campers and the professors.

If believing the same relativized proposition is not what the members of each of the groups have in common with one another, why is it being used as a principle of classification? I propose we look at things in this way. The shoppers, for example, are all in a certain belief state, a state which, given normal desires and other belief states they can be expected to be in, will lead each of them to examine his trolley. But, although they are all in the same belief state (not the same *total* belief state, of course), they do not all have the same belief (believe the same thing, have the relation of belief, to the same object).

We use sentences with indexicals or relativized propositions to individuate belief states, for the purposes of classifying believers in-ways useful for explanation and prediction. That is, belief states individuated in this way enter into our commonsense theory about human behaviour and more sophisticated theories emerging from it. We expect all good-hearted people in that state, which leads them to say "I am making a mess" to examine their grocery trolleys, no matter what belief they have in virtue of being in that state. That we individuate belief states in this way doubtless has something to do with the fact that one criterion for being in the states we postulate, at least for articulate, sincere adults, is being disposed to utter the indexical sentence in question. A good philosophy of mind should explain this in detail; my aim is merely to clarify what it is that needs explaining.

The proposal, then, is that there is not an identity, or even an isomorphic correspondence, but only a systematic relationship between the belief states one is in and what one thereby believes. The opposite assumption, that belief states should be classified by propositions believed, seems to be built right into traditional philosophies of belief. Given this assumption, whenever we have believers in the same belief state, we must expect to find a proposition they all believe, and differences in belief state lead us to expect a difference in proposition believed. The bulk of this paper consisted in following such leads to nowhere (or to propositions of limited accessibility).

Consider a believer whose belief states are characterized by a structure of sentences with indexicals or relativized propositions (those marked "true" in a very comprehensive exam, if we are dealing with an articulate, sincere adult). This structure, together with the context of belief—the time and identity of the speaker—will yield a structure of *de re* propositions. The sequence of objects will consist of the values which the indexicals take in the context. The open propositions will be those yielded by the relativized proposition when shorn of its indexical elements. These are what the person believes, in virtue of being in the states he is in, when and where he is in them.[9]

This latter structure is important, and classifications of believers by *what* they believe is appropriate for many purposes. For example, usually, when a believer moves from context to context, his belief states adjust to preserve beliefs held. As time passes, I go from the state corresponding to "The meeting will begin" to the one corresponding to "The meeting is beginning" and finally to "The meeting has begun". All along I believe of noon that it is when the meeting begins. But I believe it in different ways. And to these different ways of believing the same thing, different actions are appropriate: preparation, movement, apology. Of course if the change of context is not noted, the adjustment of belief states will not occur, and a wholesale change from believing truly to believing falsely may occur. This is what happened to Rip Van Winkle. He awakes in the same belief states he fell asleep in twenty years earlier, unadjusted to the dramatic change in context, and so with a whole new set of beliefs, such as that he is a young man, mostly false.

We have here a metaphysically benign form of limited accessibility. Anyone at any time can have access to any proposition. But not in any way. Anyone can believe of John Perry that he is making a mess. And anyone can be in the belief state classified by the sentence "I am making a mess". But only I can have that belief by being in that state.

There is room in this scheme for *de dicto* propositions, since the characterization of one's belief states may include sentences without any indexical element. If there are any, they could appear on the exam. For this part of the structure, the hypothesis of perfect correspondence would be correct.

A more radical proposal would do away with objects of belief entirely. We would think of belief as a system of relations of various degrees between persons and other objects. Rather than saying I believed in the *de re* proposition consisting of me and the open proposition *x is making a mess*, we would say that I stand in the relation, believing to be making a mess, to myself. There are many ways to stand in this relation to myself, that is, a variety of belief states I might be in. And these would be classified by sentences with indexicals. On this view *de dicto* belief, already demoted from its central place in the philosophy of belief, might be seen as merely an illusion, engendered by the implicit nature of much indexicality.

To say that belief states must be distinguished from objects of belief, cannot be individuated in terms of them, and are what is crucial for the explanation of action is not to give a full-fledged account of belief, or even a sketchy one. Similarly, to say that we must distinguish the object seen from the state of the seeing subject, and that the latter is crucial for the explanation of action guided by vision, is not to offer a full-fledged account of vision. But just as the arguments from illusion and perceptual relativity teach us that no philosophy of perception can be plausible that is not cognizant of this last distinction, the problem of the essential indexical should teach us that no philosophy of belief can be plausible that does not take account of the first.[10]

Notes

1. In thinking about the problem of the essential indexical, I have been greatly helped by the writings of Hector-Neri Castañeda on indexicality and related topics. Casteñeda focused attention on these problems, and made many of the points made here, in " 'He': A Study in the Logic of Self-consciousness", *Ratio* 8 (1966), 130–57; 'Indicators and Quasi-indicators', *American Philosophical Quarterly* 4 (1967), 85–100; and 'On the Logic of Attributions of Self Knowledge to Others', *The Journal of Philosophy* 65 (1968), 439–56. More recently his views on these matters have been developed as a part of his comprehensive system of generalized phenomenalism. See particularly 'On the Philosophical Foundations of the Theory of Communication: Reference', *Midwestern Studies in Philosophy* 2 (1977), 165–86 and 'Perception, Belief, and the Structure of Physical Objects and Consciousness', *Synthese* 35 (1977), 285–351. Having benefited so much from Castañeda's collection of 'proto-philosophical data,' I regret that differences of approach and limitations of competence and space have prevented me from incorporating a discussion of his theory into this essay. I hope to make good this omission at some future time.
2. See John Perry, 'Frege on Demonstratives', *Philosophical Review* 86 (1977), 474–97, for a critique of Frege's views on indexicality.
3. For the classic discussion of these problems, see Williard van Orman Quine, 'Quantifiers and Propositional Attitudes', reprinted in *Ways of Paradox* (New York: Random House, 1966), 183–94.
4. David Kaplan, 'Quantifying In', in Donald Davidson and Jaakko Hintikka (eds.), *Words and Objections* (Dordrecht: Reidel, 1969), 206–42.
5. Jaakko Hintikka, 'Individuals, Possible Worlds, and Epistemic Logic', *Noûs* 1 (1967), 33–62, esp. section II, 40–62.
6. See Richard Montague, 'Pragmatics', in R. Thomason (ed.), *Formal Philosophy* (Yale University Press, 1974), pp. 95–118, and Dana Scott, 'Advice on Modal Logic' in *Karel Lambert (ed.), Philosophical Problems in Logic* (Dordrecht: Reidel, 1970), 143–73, esp. the former.
7. Frege seems to accept something like it, as necessary for dealing with "I", in

'Thoughts', translated by A. M. and Marcelle Quinton, *Mind* 65 (1956), 289–311; reprinted in this volume.

8. See esp. section II of 'Perception, Belief, and the Structure of Physical Objects and Consciousness'.
9. This two-tiered structure of belief states and propositions believed will remind the reader familiar with David Kaplan's *Demonstratives, Draft #2*, mimeographed (UCLA: Department of Philosophy, 1977) of his system of characters and contents. This is no accident, for my approach to the problem of the essential indexical was formed by using the distinction as found in earlier versions of Kaplan's work to try to find a solution to the problem as articulated by Castañeda. Kaplan's treatment of indexicality was by and large shaped by considerations other than the problem of the essential indexical. So, while any plausibility one finds in what I say about that problem should be transmitted to the general outlines of his system, at least, by providing an epistemological motivation for something like the character/content distinction, any implausibility one finds will not necessarily be so transmitted. Nor should one take any details one manages to find in this essay as a guide to the details of Kaplan's system.
10. Versions of this paper were read at philosophy department colloquia at UCLA, Claremont Graduate School, and Stanford, to the Washington State University at Bellingham Philosophy Conference, and to the Meeting of Alberta Philosophy Department. I am indebted to philosophers participating in these colloquia for many helpful criticisms and comments. I owe a special debt to Michael Bratman, and Dagfinn Føllesdal, for detailed comments on the penultimate version. Most of the ideas in this paper were developed while I held a fellowship from the Guggenheim Foundation and was on sabbatical leave from Stanford University, and I thank both for their support.

VII

A PUZZLE ABOUT BELIEF

SAUL A. KRIPKE

IN this paper I will present a puzzle about names and belief. A moral or two will be drawn about some other arguments that have occasionally been advanced in this area, but my main thesis is a simple one: that the puzzle *is* a puzzle. And, as a corollary, that any account of belief must ultimately come to grips with it. Any speculation as to solutions can be deferred.

The first section of the paper gives the theoretical background in previous discussion, and in my own earlier work, that led me to consider the puzzle. The background is by no means necessary to *state* the puzzle: As a philosophical puzzle, it stands on its own, and I think its fundamental interest for the problem of belief goes beyond the background that engendered it. As I indicate in the third section, the problem really goes beyond beliefs expressed using names, to a far wider class of beliefs. Nevertheless, I think that the background illuminates the genesis of the puzzle, and it will enable me to draw one moral in the concluding section.

The second section states some general principles which underlie our general practice of reporting beliefs. These principles are stated in much more detail than is needed to comprehend the puzzle; and there are variant formulations of the principles that would do as well. Neither this section nor the first is necessary for an intuitive grasp of the central problem, discussed in the third section, though they may help with fine points of the discussion. The reader who wishes rapid access to the central problem could skim the first two sections lightly on a first reading.

In one sense the problem may strike some as no puzzle at all.

Saul Kripke, 'A Puzzle about Belief', in *Meaning and Use* (ed.) A. Margalit (Dordrecht: Reidel, 1979) 239–83.

For, in the situation to be envisaged, all the relevant facts can be described in *one* terminology without difficulty. But, in *another* terminology, the situation seems to be impossible to describe in a consistent way. This will become clearer later.

I. PRELIMINARIES: SUBSTITUTIVITY

In other writings,[1] I developed a view of proper names closer in many ways to the old Millian paradigm of naming than to the Fragean tradition which probably was dominant until recently. According to Mill, a proper name is, so to speak, *simply* a name. It *simply* refers to its bearer, and has no other linguistic function. In particular, unlike a definite description, a name does not describe its bearer as possessing any special identifying properties.

The opposing Fregean view holds that to each proper name, a speaker of the language associates some property or (conjunction of properties) which determines its referent as the unique thing fulfilling the associated property (or properties). This property(ies) constitutes the 'sense' of the name. Presumably, if '. . .' is a proper name, the associated properties are those that the speaker would supply if asked, "Who is '. . .'?" If he would answer ". . . is the man who ——", the properties filling the second blank are those that determine the reference of the name for the given speaker and constitute its 'sense.' Of course, given the name of a famous historical figure, individuals may give different, and equally correct, answers to the "Who is '. . .'?" question. Some may identify Aristotle as the philosopher who taught Alexander the Great, others as the Stagirite philosopher who studied with Plato. For these two speakers, the sense of "Aristotle" will differ: in particular, speakers of the second kind, but not of the first kind, will regard "Aristotle, if he existed, was born in Stagira" as analytic.[2] Frege (and Russell)[3] concluded that, strictly speaking, different speakers of English (or German!) ordinarily use a name such as 'Aristotle' in different senses (though with the same reference). Differences in properties associated with such names, strictly speaking, yield different idiolects.[4]

Some later theorists in the Frege–Russellian tradition have found this consequence unattractive. So they have tried to modify the view by 'clustering' the sense of the name (e.g. Aristotle is the thing having the following long list of properties, or at any rate most of them), or, better for the present purpose, socializing it (what determines the reference of 'Aristotle' is some roughly specified set of *community-wide* beliefs about Aristotle).

One way to point up the contrast between the strict Millian view and Fregean views involves—if we permit ourselves this jargon—the notion of propositional content. If a strict Millian view is correct, and the linguistic function of a proper name is completely exhausted by the fact that it names its bearer, it would appear that proper names of the same thing are everywhere interchangeable not only *salva veritate* but even *salva significatione*: the proposition expressed by a sentence should remain the same no matter what name of the object it uses. Of course this will not be true if the names are 'mentioned' rather than 'used': " 'Cicero' has six letters" differs from " 'Tully' has six letters" in truth value, let alone in content. (The example, of course, is Quine's.) Let us confine ourselves at this stage to *simple* sentences involving no connectives or other sources of intensionality. If Mill is completely right, not only should "Cicero was lazy" have the same *truth value* as "Tully was lazy", but the two sentences should express the same *proposition*, have the same content. Similarly, "Cicero admired Tully", "Tully admired Cicero", "Cicero admired Cicero", and "Tully admired Tully" should be four ways of saying the same thing.[5]

If such a consequence of Mill's view is accepted, it would seem to have further consequences regarding 'intensional' contexts. Whether a sentence expresses a necessary truth or a contingent one depends only on the proposition expressed and not on the words used to express it. So any simple sentence should retain its 'modal value' (necessary, impossible, contingently true, or contingently false) when 'Cicero' is replaced by 'Tully' in one or more places, since such a replacement leaves the content of the sentence unaltered. Of course this implies that coreferential names are substitutable in modal contexts *salva veritate*: "It is necessary (possible) that Cicero . . ." and "It is necessary (possible) that Tully . . ." must have the same truth value no matter how the dots are filled by a simple sentence.

The situation would seem to be similar with respect to contexts involving knowledge, belief, and epistemic modalities. Whether a given subject believes something is presumably true or false of such a subject no matter how that belief is expressed; so if proper name substitution does not change the content of a sentence expressing a belief, coreferential proper names should be interchangeable *salva veritate* in belief contexts. Similar reasoning would hold for epistemic contexts ("Jones knows that . . .") and contexts of epistemic necessity ("Jones knows *a priori* that . . ."), and the like.

All this, of course, would contrast strongly with the case of definite descriptions. It is well known that substitution of coreferential descriptions in simple sentences (without operators), on any reasonable conception of 'content,' *can* alter the content of such a sentence. In particular, the modal value of a sentence is not invariant under changes of coreferential descriptions: "The smallest prime is even" expresses a necessary truth, but "Jones's favourite number is even" expresses a contingent one, even if Jones's favourite number happens to be the smallest prime. It follows that coreferential descriptions are *not* interchangeable *salva veritate* in modal contexts: "It is necessary that the smallest prime is even" is true while "It is necessary that Jones's favourite number is even" is false.

Of course there is a '*de re*' or 'large scope' reading under which the second sentence is true. Such a reading would be expressed more accurately by "Jones's favourite number is such that it is necessarily even" or, in rough Russellian transcription, as "One and only one number is admired by Jones above all others, and any such number is necessarily even (has the property of necessary evenness)." Such a *de re* reading, if it makes sense at all, by definition must be subject to a principle of substitution *salva veritate*, since necessary evenness is a property of the *number*, independently of how it is designated; in this respect there can be no contrast between names and descriptions. The contrast, according to the Millian view, must come in the *de dicto* or "small scope" reading, which is the *only* reading, for belief contexts as well as modal contexts, that will concern us in this paper. If we wish, we can emphasize that this is our reading in various ways. Say, "It is necessary that: Cicero was bald" or, more explicitly, "The following proposition is necessarily true: Cicero was bald", or even, in Carnap's 'formal' mode of speech[6] " 'Cicero was bald' expresses a necessary truth". Now the Millian asserts that all these formulations retain their truth value when 'Cicero' is replaced by 'Tully,' even though 'Jones's favourite Latin author' and 'the man who denounced Catiline' would *not* similarly be interchangeable in these contexts even if they are codesignative.

Similarly for belief contexts. Here too *de re* beliefs—as in

"Jones believes, *of* Cicero (or: *of* his favourite Latin author) that he was bald" do *not* concern us in this paper. Such contexs, if they make sense, are by definition subject to a substitutivity principle for both names and descriptions. Rather we are concerned with the *de dicto* locution expressed explicitly in such formulations as, "Jones believes that: Cicero was bald" (or: "Jones believes that: the man who denounced Catiline was bald"). The material after the colon expresses the *content* of Jones's belief. Other, more explicit, formulations are: "Jones believes the proposition—that—Cicero—was—bald", or even in the 'formal' mode, "The sentence 'Cicero was bald' gives the content of a belief of Jones". In all such contexts, the strict Millian seems to be committed to saying that codesignative names, but not codesignative descriptions, are interchangeable *salva veritate*.[7]

Now it has been widely assumed that these apparent consequences of the Millian view are plainly false. First, it seemed that sentences can alter their *modal* values by replacing a name by a codesignative one. "Hesperus is Hesperus" (or, more cautiously: "If Hesperus exists, Hesperus is Hesperus") expresses a necessary truth, while "Hesperus is Phosphorus" (or: "If Hesperus exists, Hesperus is Phosphorus"), expresses an empirical discovery, and hence, it has been widely assumed, a contingent truth. (It might have turned out, and hence might have been, otherwise.)

It has seemed even more obvious that codesignative proper names are not interchangeable in belief contexts and epistemic contexts. Tom, a normal speaker of the language, may sincerely assent to "Tully denounced Catiline", but not to "Cicero denounced Catiline". He may even deny the latter. And his denial is compatible with his status as a normal English speaker who satisfies normal criteria for using both 'Cicero' and 'Tully' as names for the famed Roman (without knowing that 'Cicero' and 'Tully' name the same person). Given this, it seems obvious that Tom believes that: Tully denounced Catiline, but that he does not believe (lacks the belief) that: Cicero denounced Catiline.[8] So it seems clear that codesignative proper names are not interchangeable in belief contexts. It also seems clear that there must be two distinct propositions or contents expressed by 'Cicero denounced Catiline' and 'Tully denounced Catiline'. How else can Tom believe one and deny the other? And the difference in propositions thus expressed can only come from a difference in *sense* between 'Tully' and 'Cicero'. Such a conclusion agrees with a Fregean theory and seems to be incompatible with a purely Millian view.[9]

In the previous work mentioned above, I rejected one of these arguments against Mill, the modal argument. 'Hesperus is Phosphorus', I maintained, expresses just as necessary a truth as 'Hesperus is Hesperus'; there are no counterfactual situations in which Hesperus and Phosphorus would have been different. Admittedly, the truth of 'Hesperus is Phosphorus' was not known *a priori*, and may even have been widely disbelieved before appropriate empirical evidence came in. But these epistemic questions should be separated, I have argued, from the metaphysical question of the necessity of 'Hesperus is Phosphorus'. And it is a consequence of my conception of names as 'rigid designators' that codesignative proper names are interchangeable *salva veritate* in all contexts of (metaphysical) necessity and possibility; further, that replacement of a proper name by a codesignative name leaves the modal value of any sentence unchanged.

But although my position confirmed the Millian account of names in modal contexts, it equally appears at first blush to imply a *non-Millian* account of epistemic and belief contexts (and other contexts of propositional attitude). For I presupposed a sharp contrast between epistemic and metaphysical possibility: Before appropriate empirical discoveries were made, men might well have failed to know that Hesperus was Phosphorus, or even to believe it, even though they of course knew and believed that Hesperus was Hesperus. Does not this support a Fregean position that 'Hesperus' and 'Phosphorus' have different 'modes of presentation' that determine their references? What else can account for the fact that, before astronomers identified the two heavenly bodies, a sentence using 'Hesperus' could express a common belief, while the same context involving 'Phosphorus' did not? In the case of 'Hesperus' and 'Phosphorus', it is pretty clear what the different 'modes of presentation' would be: one mode determines a heavenly body by its typical position and appearance, in the appropriate season, in the evening; the other determines the same body by its position and appearance, in the appropriate season, in the morning. So it appears that even though, according to my view, proper names would be *modally* rigid—would have the same reference when we use them to speak of counterfactual situations as they do when used to describe the actual world—they would have a kind of Fregean 'sense' according to how that rigid reference is fixed. And the divergences of 'sense' (in this sense of 'sense') would lead to failures of interchangeability of codesignative names in contexts of propositional attitude, though not in modal contexts. Such a theory would agree with Mill regarding modal contexts but with Frege regarding belief contexts. The theory would not be *purely* Millian.[10]

After further thought, however, the Fregean conclusion appears less obvious. Just as people are said to have been unaware at one time of the fact that Hesperus is Phosphorus, so a normal speaker of English apparently may not know that Cicero is Tully, or that Holland is the Netherlands. For he may sincerely assent to 'Cicero was lazy', while dissenting from 'Tully was lazy', or he may sincerely assent to 'Holland is a beautiful country', while dissenting from 'The Netherlands is a beautiful country'. In the case of 'Hesperus' and 'Phosphorus', it seemed plausible to account for the parallel situation by supposing that 'Hesperus' and 'Phosphorus' fixed their (rigid) references to a single object in two conventionally different ways, one as the 'evening star' and one as the 'morning star'. But what corresponding *conventional* 'senses,' even taking 'senses' to be 'modes of fixing the reference rigidly', can plausibly be supposed to exist for 'Cicero' and 'Tully' (or 'Holland' and 'the Netherlands')? Are not these just two names (in English) for the same man? Is there any special *conventional*, *community-wide* 'connotation' in the one lacking in the other?[11] I am unaware of any.[12]

Such considerations might seem to push us toward the extreme Frege–Russellian view that the senses of proper names vary, strictly speaking, from speaker to speaker, and that there is no community-wide sense but only a community-wide reference.[13] According to such a view, the sense a given speaker attributes to such a name as 'Cicero' depends on which assertions beginning with 'Cicero' he accepts and which of these he regards as *defining*, for him, the name (as opposed to those he regards as mere factual beliefs 'about Cicero'). Similarly, for 'Tully'. For example, someone may define 'Cicero' as 'the Roman orator whose speech was Greek to Cassius', and 'Tully' as 'the Roman orator who denounced Catiline'. Then such a speaker may well fail to accept 'Cicero is Tully' if he is unaware that a single orator satisfied both descriptions (if Shakespeare and history are both to be believed). He may well, in his ignorance, affirm 'Cicero was bald' while rejecting 'Tully was bald', and the like. Is this not what actually occurs whenever someone's expressed beliefs fail to be indifferent to interchange of 'Tully' and 'Cicero'? Must not the source of such a failure lie in two distinct associated descriptions, or modes of determining the reference, of the two names? If a speaker does, as luck would have it, attach the same identifying properties both to 'Cicero' and to 'Tully,' he *will*, it would seem, use 'Cicero' and 'Tully' interchangeably. All this appears at first blush to be powerful support for the view of Frege and Russell that in general names are peculiar to idiolects, with 'senses' depending on the associated 'identifying descriptions'.

Note that, according to the view we are now entertaining, one *cannot* say, "Some people are unaware that Cicero is Tully." For, according to this view, there is no single proposition denoted by the 'that' clause, that the community of normal English speakers expresses by 'Cicero is Tully'. Some—for example, those who define both 'Cicero' and 'Tully' as 'the author of *De Fato*'—use it to express a trivial self-identity. Others use it to express the proposition that the man who satisfied one description (say, that he denounced Catiline) is one and the same as the man who satisfied another (say, that his speech was Greek to Cassius). There is no single fact, 'that Cicero is Tully', known by some but not all members of the community.

If I were to assert, "Many are unaware that Cicero is Tully", *I* would use 'that Cicero is Tully' to denote the proposition that *I* understand by these words. If this, for example, is a trivial self-identity, I would assert falsely, and irrelevantly, that there is widespread ignorance in the community of a certain self-identity.[14] I *can*, of course, say, "Some English speakers use both 'Cicero' and 'Tully' with the usual referent (the famed Roman) yet do not assent to 'Cicero is Tully.'"

This aspect of the Frege–Russellian view can, as before, be combined with a concession that names are rigid designators and that hence the description used to fix the reference of a name is not synonymous with it. But there are considerable difficulties. There is the obvious intuitive unpalatability of the notion that we use such proper names as 'Cicero', 'Venice', 'Venus' (the planet) with differing 'senses' and for this reason do not 'strictly speaking' speak a single language. There are the many well-known and weighty objections to any description or cluster-of-descriptions theory of names. And is it definitely so clear that failure of interchangeability in belief contexts implies some difference of sense? After all, there is a considerable philosophical literature arguing that even word pairs that are straightforward synonyms if any pairs are—"doctor" and "physician," to give one example—are not interchangeable *salva veritate* in belief contexts, at least if the belief operators are iterated.[15]

A minor problem with this presentation of the argument for Frege and Russell will emerge in the next section: if Frege and Russell are right, it is not easy to state the very argument from belief contexts that appears to support them.

But the clearest objection, which shows that the others should be given their proper weight, is this: the view under consideration does not in fact account for the phenomena it seeks to explain. As I have said elsewhere,[16] individuals who "define 'Cicero' " by such phrases as "the Catiline denouncer", "the author of *De Fato*", etc. are relatively rare: their prevalence in the philosophical literature is the product of the excessive classical learning of some philosophers. Common men who clearly use 'Cicero' as a name for Cicero may be able to give no better answer to "Who was Cicero?" than "a famous Roman orator", and they probably would say the same (if anything!) for 'Tully'. (Actually, most people probably have never heard the name 'Tully'.) Similarly, many people who have heard of both Feynman and Gell-Mann would identify each as 'a leading contemporary theoretical physicist'. Such people do not assign 'senses' of the usual type to the names that uniquely identify the referent (even though they use the names with a determinate reference). But to the extent that the *indefinite* descriptions attached or associated can be called 'senses', the 'senses' assigned to 'Cicero' and 'Tully', or to 'Feynman' and 'Gell-Mann', are *identical*.[17] Yet clearly speakers of this type can ask, "Were Cicero and Tully one Roman orator, or two different ones?" or "Are Feynman and Gell-Mann two different physicists, or one?" without knowing the answer to either question by inspecting 'senses' alone. Some such speaker might even conjecture, or be under the vague false impression, that, as he would say, 'Cicero was bald but Tully was not'. The premiss of the argument we are considering for the classic position of Frege and Russell—that whenever two codesignative names fail to be interchangeable in the expression of a speaker's beliefs, failure of interchangeability arises from a difference in the 'defining' descriptions the speaker associates with these names—is, therefore, false. The case illustrated by 'Cicero' and 'Tully' is, in fact, quite usual and ordinary. So the apparent failure of codesignative names to be everywhere interchangeable in belief contexts is not to be explained by differences in the 'senses' of these names.

Since the extreme view of Frege and Russell does not in fact explain the apparent failure of the interchangeability of names in belief contexts, there seems to be no further reason—for present purposes—not to give the other overwhelming *prima facie* considerations against the Frege–Russell view their full weight. Names of famous cities, countries, persons, and planets are the common currency of our common language, not terms used homonymously in our separate idiolects.[18] The apparent failure of codesignative names to be interchangeable in belief contexts remains a mystery, but the mystery no longer seems so clearly to argue for a Fregean view as against a Millian one. Neither differing public senses nor differing private senses peculiar to each speaker account for the phenomena to be explained. So the apparent existence of such phenomena no longer gives a *prima facie* argument for such differing senses.

One final remark to close this section. I have referred before to my own earlier views in "Naming and Necessity". I said above that these views, inasmuch as they make proper names rigid and transparent[19] in modal contexts, favour Mill, but that the concession that proper names are not transparent in belief contexts appears to favour Frege. On a closer examination, however, the extent to which these opacity phenomena really support Frege against Mill becomes much more doubtful. And there are important theoretical reasons for viewing the 'Naming and Necessity' approach in a Millian light. In that work I argued that ordinarily the real determinant of the reference of names of a former historical figure is a chain of communication, in which the reference of the name is passed from link to link. Now the legitimacy of such a chain accords much more with Millian views than with alternatives. For the view supposes that a learner acquires a name from the community by determining to use it with the same reference as does the community. We regard such a learner as using "Cicero is bald" to express the same thing the community expresses, regardless of variations in the properties different learners associate with 'Cicero', as long as he determines that he will use the name with the referent current in the community. That a name can be transmitted in this way accords nicely with a Millian picture, according to which only the reference, not more specific properties associated with the name, is relevant to the semantics of sentences containing it. It has been suggested that the chain of communication, which on the present picture determines the reference, might thereby itself be called a 'sense'. Perhaps so—if we wish[20]—but we should not thereby forget that the legitimacy of such a chain suggests that it is just preservation of reference, as Mill thought, that we regard as necessary for correct language learning.[21] (This contrasts with such terms as 'renate' and 'cordate', where more than learning the correct extension is needed.) Also, as suggested above, the doctrine of rigidity in modal contexts is dissonant, though not necessarily inconsistent, with a view that invokes anti-Millian considerations to explain propositional attitude contexts.

The spirit of my earlier views, then, suggests that a Millian line should be maintained as far as is feasible.

II. PRELIMINARIES: SOME GENERAL PRINCIPLES

Where are we now? We seem to be in something of a quandary. On the one hand, we concluded that the failure of 'Cicero' and 'Tully' to be interchangeable *salva veritate* in contexts of propositional attitude was by no means explicable in terms of different 'senses' of the two names. On the other hand, let us not forget the initial argument against Mill: If reference is *all there is* to naming, what semantic difference can there be between 'Cicero' and 'Tully'? And if there is no semantic difference, do not 'Cicero was bald' and 'Tully was bald' express exactly the same proposition? How, then, can anyone believe that Cicero was bald, yet doubt or disbelieve that Tully was?

Let us take stock. Why do we think that anyone can believe that Cicero was bald, but fail to believe that Tully was? Or believe, without any logical inconsistency, that Yale is a fine university, but that Old Eli is an inferior one? Well, a normal English speaker, Jones, can sincerely assent to 'Cicero was bald' but not to 'Tully was bald'. And this even though Jones uses 'Cicero' and 'Tully' in standard ways—he uses 'Cicero' in this assertion as a name for the Roman, not, say, for his dog, or for a German spy.

Let us make explicit the *disquotational principle* presupposed here, connecting sincere assent and belief. It can be stated as follows, where '*p*' is to be replaced, inside and outside all quotation marks, by any appropriate standard English sentence: "*If a normal English speaker, on reflection, sincerely assents to 'p', then he believes that p*." The sentence replacing '*p*' is to lack indexical or pronominal devices or ambiguities that would ruin the intuitive sense of the principle (e.g. if he assents to "You are wonderful", he need not believe that *you*—the reader—are wonderful).[22] When we suppose that we are dealing with a normal speaker of English, we mean that he uses all words in the sentence in a standard way, combines them according to the appropriate syntax, etc.: in short, he uses the sentence to mean what a normal speaker should mean by it. The 'words' of the sentence may include proper names, where these are part of the common discourse of the community, so that we can speak of using them in a standard way. For example, if the sentence is "London is pretty", then the speaker should satisfy normal criteria for using 'London' as a name of London, and for using 'is pretty' to attribute an appropriate degree of pulchritude. The qualification "on reflection" guards against the possibility that a speaker may, through careless inattention to the meaning of his words or other momentary conceptual or linguistic confusion, assert something he does not really mean, or assent to a sentence in linguistic error. "Sincerely" is meant to exclude mendacity, acting, irony, and the like. I fear that even with all this it is possible that some astute reader—such, after all, is the way of philosophy—may discover a qualification I have overlooked, without which the asserted principle is subject to counter-example. I doubt, however, that any such modification will affect any of the uses of the principle to be considered below. Taken in its obvious intent, after all, the principle appears to be a self-evident truth. (A similar principle holds for sincere affirmation or assertion in place of assent.)

There is also a strengthened 'biconditional' form of the disquotational principle, where once again any appropriate English sentence may replace '*p*' throughout: *A normal English speaker who is not reticent will be disposed to sincere reflective assent to 'p' if and only if he believes that p*.[23] The biconditional form strengthens the simple one by adding that failure to assent indicates lack of belief, as assent indicates belief. The qualification about reticence is meant to take account of the fact that a speaker may fail to avow his beliefs because of shyness, a desire for secrecy, to avoid offence, etc. (An alternative formulation would give the speaker a sign to indicate lack of belief—not necessarily disbelief—in the assertion propounded, in addition to his sign of assent.) Maybe again the formulation needs further tightening, but the intent is clear.

Usually below, the simple disquotational principle will be sufficient for our purposes, but once we will also invoke the strengthened form. The simple form can often be used as a test for disbelief, provided the subject is a speaker with the modicum of logicality needed so that, at least after appropriate reflection, he does not hold simultaneously beliefs that are straightforward contradictions of each other—of the forms '*p*' and '$\sim p$'.[24] (Nothing in such a requirement prevents him from holding simultaneous beliefs that jointly *entail* a contradiction.) In this case (where '*p*' may be replaced by any appropriate English sentence), the speaker's assent to the negation of '*p*' indicates not only his disbelief that *p* but also his failure to believe that *p*, using only the simple (unstrengthened) disquotational principle.

So far our principle applies only to speakers of English. It allows us to infer, from Peter's sincere reflective assent to "God exists", that he believes that God exists. But of course we ordinarily allow ourselves to draw conclusions, stated in English, about the beliefs of speakers of any language: we infer that Pierre believes that God exists from his sincere reflective assent to "*Dieu existe*". There are several ways to do this, given conventional translations of French into English. We choose the following route. We have stated the disquotational principle in English, for English sentences; an analogous principle, stated in French (German, etc.) will be assumed to hold for French (German, etc.) sentences. Finally, we assume the *principle of translation: If a sentence of one language expresses a truth in that language, then any translation of it into any other language also expresses a truth (in that other language)*. Some of our ordinary practice of translation may violate this principle; this happens when the translator's aim is not to preserve the content of the sentence, but to serve—in some other sense—the same purposes in the home language as the original utterance served in the foreign language.[25] But if the translation of a sentence is to mean the same as the sentence translated, preservation of truth value is a minimal condition that must be observed.

Granted the disquotational principle expressed in each language, reasoning starting from Pierre's assent to '*Dieu existe*' continues thus. First, on the basis of his utterance and the French disquotational principle we infer (in French):

Pierre croit que Dieu existe.

From this we deduce,[26] using the principle of translation:

Pierre believes that God exists.

In this way we can apply the disquotational technique to all languages.

Even if I apply the disquotational technique to English alone, there is a sense in which I can be regarded as tacitly invoking a principle of translation. For presumably I apply it to speakers of the language other than myself. As Quine has pointed out, to regard others as speaking the same language as I is in a sense tacitly to assume a *homophonic* translation of their language into my own. So when I infer from Peter's sincere assent to or affirmation of "God exists" that he believes that God exists, it is arguable that, strictly speaking, I combine the disquotational principle (for Peter's idiolect) with the principle of (homophonic) translation (of Peter's idiolect into mine). But for most purposes, we can formulate the disquotational principle for a single language, English, tacitly supposed to be the common language of English speakers. Only when the possibility of individual differences of dialect is relevant need we view the matter more elaborately.

Let us return from these abstractions to our main theme. Since a normal speaker—normal even in his use of 'Cicero' and 'Tully' as names—can give sincere and reflective assent to "Cicero was bald" and simultaneously to "Tully was not bald", the disquotational principle implies that he believes that Cicero was bald and believes that Tully was not bald. Since it seems that he need not have contradictory beliefs (even if he is a brilliant logician, he need not be able to deduce that at least one of his beliefs must be in error), and since a substitutivity principle for coreferential proper names in belief contexts would imply that he does have contradictory beliefs, it would seem that such a substitutivity principle must be incorrect. Indeed, the argument appears to be a *reductio ad absurdum* of the substitutivity principle in question.

The relation of this argument against substitutivity to the classical position of Russell and Frege is a curious one. As we have seen, the argument can be used to give *prima facie* support for the Frege–Russell view, and I think many philosophers have regarded it as such support. But in fact this very argument, which has been used to support Frege and Russell, cannot be stated in a straightforward fashion if Frege and Russell are right. For suppose

Jones asserts, "Cicero was bald, but Tully was not". If Frege and Russell are right, I cannot deduce, using the disquotational principle:

(1) Jones believes that Cicero was bald but Tully was not,

since, in general, Jones and I will not, strictly speaking, share a common idiolect unless we assign the same 'senses' to all names. Nor can I combine disquotation and translation to the appropriate effect, since homophonic translation of Jones's sentence into mine will in general be incorrect for the same reason. Since in fact I make no special distinction in sense between 'Cicero' and 'Tully'—to me, and probably to you as well, these are interchangeable names for the same man—and since, according to Frege and Russell, Jones's very affirmation of (1) shows that for him there *is* some distinction of sense, Jones must therefore, on Frege–Russellian views, use one of these names differently from me, and homophonic translation is illegitimate. Hence, if Frege and Russell are right, we *cannot* use this example in the usual straightforward way to conclude that proper names are not substitutable in belief contexts—even though the example, and the ensuing negative verdict on substitutivity, has often been thought to support Frege and Russell!

Even according to the Frege–Russellian view, however, *Jones* can conclude, using the disquotational principle, and expressing his conclusion in his own idiolect:

(2) I believe that Cicero was bald but Tully was not.

I cannot endorse this conclusion in Jones's own words, since I do not share Jones's idiolect. I *can* of course conclude, "(2) expresses a truth in Jones's idiolect". I can also, if I find out the two 'senses' Jones assigns to 'Cicero' and 'Tully', introduce two names '*X*' and '*Y*' into my own language with these same two senses ('Cicero' and 'Tully' have already been preempted) and conclude:

(3) Jones believes that *X* was bald and *Y* was not.

All this is enough so that we can still conclude, on the Frege–Russellian view, that codesignative names are not interchangeable in belief contexts. Indeed this can be shown more simply on this view, since codesignative descriptions plainly are not interchangeable in these contexts, and for Frege and Russell names, being essentially abbreviated descriptions, cannot differ in this respect. Nevertheless, the simple argument, apparently free of such special Frege–Russellian doctrinal premisses (and often used to support these premisses), in fact cannot go through if Frege and Russell are right.

However, if, *pace* Frege and Russell, widely used names are common currency of our language, then there no longer is any problem for the simple argument, using the disquotational principle, to (2). So, it appears, on pain of convicting Jones of inconsistent beliefs—surely an unjust verdict—we must not hold a substitutivity principle for names in belief contexts. If we used the *strengthened* disquotational principle, we could invoke Jones's presumed lack of any tendency to assent to 'Tully was bald' to conclude that he does not believe (lacks the belief) that Tully was bald. Now the refutation of the substitutivity principle is even stronger, for when applied to the conclusion that Jones believes that Cicero was bald but does not believe that Tully was bald, it would lead to a straightout contradiction. The contradiction would no longer be in Jones's beliefs but in our own.

This reasoning, I think, has been widely accepted as proof that codesignative proper names are not interchangeable in belief contexts. Usually the reasoning is left tacit, and it may well be thought that I have made heavy weather of an obvious conclusion. I wish, however, to question the reasoning. I shall do so without challenging any particular step of the argument. Rather I shall present—and this will form the core of the present paper—an argument for a paradox about names in belief contexts that invokes *no* principle of substitutivity. Instead it will be based on the principles—apparently so obvious that their use in these arguments is ordinarily tacit—of disquotation and translation.

Usually the argument will involve more than one language, so that the principle of translation and our conventional manual of translation must be invoked. We will also give an example, however, to show that a form of the paradox may result within English alone, so that the only principle invoked is that of disquotation (or, perhaps, disquotation plus *homophonic* translation). It will intuitively be fairly clear, in these cases, that the situation of the subject is 'essentially the same' as that of Jones with respect to 'Cicero' and 'Tully'. Moreover, the paradoxical conclusions about the subject will parallel those drawn about Jones on the basis of the substitutivity principle, and the arguments will parallel those regarding Jones. Only in these cases, no special substitutivity principle is invoked.

The usual use of Jones's case as a counter-example to the substitutivity principle is thus, I think, somewhat analogous to the following sort of procedure. Someone wishes to give a *reductio ad absurdum* argument against a hypothesis in topology. He does succeed in refuting this hypothesis, but his derivation of an absurdity from the hypothesis makes essential use of the unrestricted comprehension schema in set theory, which he regards as self-evident. (In particular, the class of all classes not members of themselves plays a key role in his argument.) Once we know that the unrestricted comprehension schema and the Russell class lead to contradiction by themselves, it is clear that it was an error to blame the earlier contradiction on the topological hypothesis.

The situation would have been the same if, after deducing a contradiction from the topological hypothesis plus the 'obvious' unrestricted comprehension schema, it was found that a similar contradiction followed if we replaced the topological hypothesis by an apparently 'obvious' premiss. In both cases it would be clear that, even though we may still not be confident of any specific flaw in the argument against the topological hypothesis, blaming the contradiction on that hypothesis is illegitimate: rather we are in a 'paradoxical' area where it is unclear *what* has gone wrong.[27]

It is my suggestion, then, that the situation with respect to the interchangeability of codesignative names is similar. True, such a principle, when combined with our normal disquotational judgments of belief, leads to straightforward absurdities. But we will see that the 'same' absurdities can be derived by replacing the interchangeability principle with our normal practices of translation and disquotation, or even with disquotation alone.

The particular principle stated here gives just one particular way of 'formalizing' our normal inferences from explicit affirmation or assent to belief; other ways of doing it are possible. It is undeniable that we *do* infer, from a normal Englishman's sincere affirmation of 'God exists' or 'London is pretty', that he believes, respectively, that God exists or that London is pretty; and that we would make the same inferences from a Frenchman's affirmation of '*Dieu existe*' or '*Londres est jolie*'. Any principles that would justify such inferences are sufficient for the next section. It will be clear that the particular principles stated in the present section are sufficient, but in the next section the problem will be presented informally in terms of our inferences from foreign or domestic assertion to belief.

III. THE PUZZLE

Here, finally(!), is the puzzle. Suppose Pierre is a normal French speaker who lives in France and speaks not a word of English or of any other language except French. Of course he has heard of that famous distant city, London (which he of course calls '*Londres*'), though he himself has never left France. On the basis of what he has heard of London, he is inclined to think that it is pretty. So he says, in French, "*Londres est jolie*".

On the basis of his sincere French utterance, we will conclude:

(4) Pierre believes that London is pretty.

I am supposing that Pierre satisfies all criteria for being a normal French speaker, in particular, that he satisfies whatever criteria we usually use to judge that a Frenchman (correctly) uses '*est jolie*' to attribute pulchritude and uses '*Londres*'—standardly—as a name of London.

Later, Pierre, through fortunate or unfortunate vicissitudes, moves to England, in fact to London itself, though to an unattractive part of the city with fairly uneducated inhabitants. He, like most of his neighbours, rarely ever leaves this part of the city. None of his neighbours know any French, so he must learn English by 'direct method', without using any translation of English into French: by talking and mixing with the people he eventually begins to pick up English. In particular, everyone speaks of the city, 'London', where they all live. Let us suppose for the moment—though we will see below that this is not crucial—that the local population are so uneducated that they know few of the facts that Pierre heard about London in France. Pierre learns from them everything they know about London, but there is little overlap with what he heard before. He learns, of course—speaking English—to call the city he lives in 'London'. Pierre's surroundings are, as I said, unattractive, and he is unimpressed with most of the rest of what he happens to see. So he is inclined to assent to the English sentence:

(5) London is not pretty.

He has *no* inclination to assent to:

(6) London is pretty.

Of course he does not for a moment withdraw his assent from the French sentence, "*Londres est jolie*"; he merely takes it for granted that the ugly city in which he is now stuck is distinct from the enchanting city he heard about in France. But he has no inclination to change his mind for a moment about the city he still calls '*Londres*'.

This, then, is the puzzle. If we consider Pierre's past background as a French speaker, his entire linguistic behaviour, on the same basis as we would draw such a conclusion about many of his countrymen, supports the conclusion ((4) above) that he believes that London is pretty. On the other hand, after Pierre lived in London for some time, he did not differ from his neighbours—his French background aside—either in his knowledge of English or in his command of the relevant facts of local geography. His English vocabulary differs little from that of his neighbours. He, like them, rarely ventures from the dismal quarter of the city in which they all live. He, like them, knows that the city he lives in is called 'London' and knows a few other facts. Now Pierre's neighbours would surely be said to use 'London' as a name for London and to speak English. Since, as an English speaker, he does not differ at all from them, we should say the same of him. But then, on the basis of his sincere assent to (5), we should conclude:

(7) Pierre believes that London is not pretty.

How can we describe this situation? It seems undeniable that Pierre *once* believed that London is pretty—at least before he learnt English. For at that time, he differed not at all from countless numbers of his countrymen, and we would have exactly the same grounds to say of him, as of any of them, that he believes that London is pretty: if any Frenchman who was both ignorant of English and never visited London believed that London is pretty, Pierre did. Nor does it have any plausibility to suppose, because of his later situation *after* he learns English, that Pierre should *retroactively* be judged *never* to have believed that London is pretty. To allow such *ex post facto* legislation would, as long as the future is uncertain, endanger our attributions of belief to *all* monolingual Frenchmen. We would be forced to say that Marie, a monolingual who firmly and sincerely asserts, "*Londres est jolie*", may or may not believe that London is pretty depending on the *later* vicissitudes of her career (if later she learns English and . . .). No: Pierre, like Marie, believed that London is pretty when he was monolingual.

Should we say that Pierre, now that he lives in London and speaks English, no longer believes that London is pretty? Well, unquestionably Pierre *once* believed that London is pretty. So we would be forced to say that Pierre has *changed his mind, has given up his previous belief*. But has he really done so? Pierre is very set in his ways. He reiterates, with vigour, every assertion he has ever made in French. He says he has not changed his mind about anything, has *not* given up any belief. Can we say he is wrong about this? If we did not have the story of his living in London and his English utterances, on the basis of his normal command of French we would be *forced* to conclude that he *still* believes that London is pretty. And it does seem that this is correct. Pierre has neither changed his mind nor given up any belief he had in France.

Similar difficulties beset any attempt to deny him his new belief. His French past aside, he is just like his friends in London. Anyone else, growing up in London with the same knowledge and beliefs that he expresses in England, we would undoubtedly judge to believe that London is not pretty. Can Pierre's French past nullify such a judgement? Can we say that Pierre, because of his French past, does not believe that (5)? Suppose an electric shock wiped out all his memories of the French language, what he learnt in France, and his French past. He would then be *exactly* like his neighbours in London. He would have the *same* knowledge, beliefs, and linguistic capacities. We then presumably would be forced to say that Pierre believes that London is ugly if we say it of his neighbours. But surely no shock that *destroys* part of Pierre's memories and knowledge can *give* him a new belief. If Pierre believes (5) *after* the shock, he believed it before, despite his French language and background.

If we would deny Pierre, in his bilingual stage, his belief that London is pretty *and* his belief that London is not pretty, we combine the difficulties of both previous options. We still would be forced to judge that Pierre once believed that London is pretty but does no longer, in spite of Pierre's own sincere denial that he has lost any belief. We also must worry whether Pierre would *gain* the belief that London is not pretty if he totally forgot his French past. The option does not seem very satisfactory.

So now it seems that we must respect both Pierre's French utterances and their English counterparts. So we must say that Pierre has contradictory beliefs, that he believes that London is pretty *and* he believes that London is not pretty. But there seem to be insuperable difficulties with this alternative as well. We may suppose that Pierre, in spite of the unfortunate situation in which he now finds himself, is a leading philosopher and logician. He would *never* let contradictory beliefs pass. And surely anyone, leading logician or no, is in principle in a position to notice and correct contradictory beliefs if he has them. Precisely for this reason, we regard individuals who contradict themselves as subject to greater censure than those who merely have false beliefs. But it is clear that Pierre, as long as he is unaware that the cities he calls 'London' and '*Londres*' are one and the same, is in no position to see, by logic alone, that at least one of his beliefs must be false. He lacks information, not logical acumen. He cannot be convicted of inconsistency: to do so is incorrect.

We can shed more light on this if we change the case. Suppose that in France Pierre, instead of affirming "*Londres est jolie*", had affirmed, more cautiously, "*Si New York est jolie, Londres est jolie aussi*", so that he believed that *if* New York is pretty, so is London. Later Pierre moves to London, learns English as before, and says (in English) "London is not pretty". So he now believes, further, that London is *not* pretty. Now from the two premisses, both of which appear to be among his beliefs ((*a*) If New York is pretty, London is, and (*b*) London is not pretty), Pierre should be able to deduce by *modus tollens* that New York is not pretty. But no matter how great Pierre's logical acumen may be, *he cannot in fact make any such deduction as long as he supposes that 'Londres' and 'London' may name two different cities*. If he *did* draw such a conclusion, he would be guilty of a fallacy.

Intuitively, he may well suspect that New York is pretty, and just this suspicion may lead him to suppose that '*Londres*' and 'London' probably name distinct cities. Yet, if we follow our normal practice of reporting the beliefs of French and English speakers, *Pierre has available to him (among his beliefs) both the premisses of a* modus tollens *argument that New York is not pretty*.

Again, we may emphasize Pierre's *lack* of belief instead of his belief. Pierre, as I said, has no disposition to assent to (6). Let us concentrate on this, ignoring his disposition to assent to (5). In fact, if we wish we may change the case: Suppose Pierre's neighbours think that since they rarely venture outside their own ugly section, they have no right to any opinion as to the pulchritude of the whole city. Suppose Pierre shares their attitude. Then, judging by his failure to respond affirmatively to "London is pretty," we may judge, from Pierre's behaviour as an *English* speaker, that he lacks the belief that London is pretty: never mind whether he disbelieves it, as before, or whether, as in the modified story, he insists that he has no firm opinion on the matter.

Now (using the *strengthened* disquotational principle), we can derive a contradiction, not merely in Pierre's judgements, but in our own. For on the basis of his behaviour as an English speaker, we concluded that he does *not* believe that London is pretty (that is, that it is not the case that he believes that London is pretty). But on the basis of his behaviour as a *French* speaker, we must conclude that he *does* believe that London is pretty. This is a contradiction.[28]

We have examined four possibilities for characterizing Pierre while he is in London: (*a*) that at that time we no longer respect his French utterance ('*Londres est jolie*'), that is that we no longer ascribe to him the corresponding belief; (*b*) that we do not respect his English utterance (or lack of utterance); (*c*) that we respect neither; (*d*) that we respect both. Each possibility seems to lead us to say something either plainly false or even downright contradictory. Yet the possibilities appear to be logically exhaustive. This, then, is the paradox.

I have no firm belief as to how to solve it. But beware of one source of confusion. It is no solution in itself to observe that some *other* terminology, which evades the question whether Pierre believes that London is pretty, may be sufficient to state all the relevant facts. I am fully aware that complete and straightforward descriptions of the situation are possible and that in this sense there is no paradox. Pierre is disposed to sincere assent to '*Londres est jolie*' but not to 'London is pretty'. He uses French normally, English normally. Both with '*Londres*' and 'London' he associates properties sufficient to determine that famous city, but he does not realize that they determine a single city. (And his uses of '*Londres*' and 'London' are historically (causally) connected with the same single city, though he is unaware of that.) We may even give a rough statement of his beliefs. He believes that the city he calls '*Londres*' is pretty, that the city he calls 'London' is not.

No doubt other straightforward descriptions are possible. No doubt some of these are, in a certain sense, *complete* descriptions of the situation.

But none of this answers the original question. Does Pierre, or does he not, believe that London is pretty? I know of no answer to *this* question that seems satisfactory. It is no answer to protest that, in some *other* terminology, one can state 'all the relevant facts'.

To reiterate, this is the puzzle: Does Pierre, or does he not, believe that London is pretty? It is clear that our normal criteria for the attribution of belief lead, when applied to *this* question, to paradoxes and contradictions. One set of principles adequate to many ordinary attributions of belief, but which leads to paradox in the present case, was stated in Section II; and other formulations are possible. As in the case of the logical paradoxes, the present puzzle presents us with a problem for customarily accepted principles and a challenge to formulate an acceptable set of principles that does not lead to paradox, is intuitively sound, and supports the inferences we usually make. Such a challenge cannot be met simply by a description of Pierre's situation that evades the question whether he believes that London is pretty.

One aspect of the presentation may misleadingly suggest the applicability of Frege–Russellian ideas that each speaker associates his own description or properties to each name. For as I just set up the case Pierre learnt one set of facts about the so-called '*Londres*' when he was in France, and *another* set of facts about 'London' in England. Thus it may appear that 'what is really going on' is that Pierre believes that *the city* satisfying *one* set of properties *is* pretty, while he believes that *the city* satisfying *another* set of properties *is not* pretty.

As we just emphasized, the phrase 'what is really going on' is a danger signal in discussions of the present paradox. The conditions stated may—let us concede for the moment—describe 'what is really going on'. But they do not resolve the problem with which we began, that of the behaviour of names in belief contexts: Does Pierre, or does he not, believe that London (not the city satisfying such-and-such descriptions, but *London*) is pretty? No answer has yet been given.

Nevertheless, these considerations may appear to indicate that descriptions, or associated properties, are highly relevant somehow to an ultimate solution, since at this stage it appears that the entire puzzle arises from the fact that Pierre originally associated different identifying properties with 'London' and '*Londres*'. Such a reaction may have some force even in the face of the now fairly well-known arguments against 'identifying descriptions' as in any way 'defining', or even 'fixing the reference', of names. But in fact the special features of the case, as I set it out, are misleading. The puzzle can arise even if Pierre associates exactly the same identifying properties with both names.

First, the considerations mentioned above in connection with 'Cicero' and 'Tully' establish this fact. For example, Pierre may well learn, in France, '*Platon*' as the name of a major Greek philosopher, and later, in England, learn 'Plato' with the same identification. Then the same puzzle can arise: Pierre may have believed, when he was in France and was monolingual in French, that Plato was bald (he would have said, "*Platon était chauve*"), and later conjecture, in English, "Plato was not bald", thus indicating that he believes or suspects that Plato was *not* bald. He need only suppose that, in spite of the similarity of their names, the man he calls '*Platon*' and the man he calls 'Plato' were two distinct major Greek philosophers. In principle, the same thing could happen with 'London' and '*Londres*'.

Of course, most of us learn a *definite* description about London, say 'the largest city in England'. Can the puzzle still arise? It is noteworthy that the puzzle can still arise even if Pierre associates to '*Londres*' and to 'London' *exactly* the same *uniquely identifying* properties. How can this be? Well, suppose that Pierre believes that London is the largest city in (and capital of) England, that it contains Buckingham Palace, the residence of the Queen of England, and he believes (correctly) that these properties, conjointly, uniquely identify the city. (In this case, it is best to suppose that he has never seen London, or even England, so that he uses *only* these properties to identify the city. Nevertheless, he has learnt English by 'direct method'.) These uniquely identifying properties he comes to associate with 'London' after he learnt English, and he expresses the appropriate beliefs about 'London' in English. Earlier, when he spoke nothing but French, however, he associated *exactly* the same uniquely identifying properties with '*Londres*'. He believed that '*Londres*', as he called it, could be uniquely identified as the capital of England, that it contained Buckingham Palace, that the Queen of England lived there, etc. Of course he expressed these beliefs, like most monolingual

Frenchmen, in French. In particular, he used '*Angleterre*' for England, '*le Palais de Buckingham*' (pronounced '*Bookeengam*'!) for Buckingham Palace, and '*la Reine d'Angleterre*' for the Queen of England. But if any Frenchman who speaks no English can ever be said to associate *exactly* the properties of being the capital of England etc. with the name '*Londres*', Pierre in his monolingual period did so.

When Pierre becomes a bilingual, *must* he conclude that 'London' and '*Londres*' name the same city, because he defined each by the same uniquely identifying properties?

Surprisingly, no! Suppose Pierre had affirmed, '*Londres est jolie*'. If Pierre has any reason—even just a 'feeling in his bones', or perhaps exposure to a photograph of a miserable area which he was told (in English) was part of 'London'—to maintain 'London is not pretty', he need not contradict himself. He need only conclude that 'England' and '*Angleterre*' name two different countries, that 'Buckingham Palace' and '*le Palais de Buckingham*' (recall the pronunciation!), name two different palaces, and so on. Then he can maintain *both* views without contradiction, and regard *both* properties as uniquely identifying.

The fact is that the paradox reproduces itself on the level of the 'uniquely identifying properties' that description theorists have regarded as 'defining' proper names (and *a fortiori*, as fixing their references). Nothing is more reasonable than to suppose that if two names, *A* and *B*, and a single set of properties, *S*, are such that a certain speaker believes that the referent of *A* uniquely satisfies all of *S* and that the referent of *B* also uniquely satisfies all of *S*, then that speaker is committed to the belief that *A* and *B* have the same reference. In fact, the identity of the referents of *A* and *B* is an easy *logical consequence* of the speaker's beliefs.

From this fact description theorists concluded that names can be regarded as synonymous, and hence interchangeable *salva veritate* even in belief contexts, provided that they are 'defined' by the same uniquely identifying properties.

We have already seen that there is a difficulty in that the set *S* of properties need not in fact be uniquely identifying. But in the present paradoxical situation there is a surprising difficulty even if the supposition of the description theorist (that the speaker believes that *S* is uniquely fulfilled) in fact holds. For, as we have seen above, Pierre is in no position to draw ordinary logical consequences from the conjoint set of what, when we consider him separately as a speaker of English and as a speaker of French, we would call his beliefs. He cannot infer a contradiction from his separate beliefs that London is pretty and that London is not pretty. Nor, in the modified situation above, would Pierre make a normal *modus tollens* inference from his beliefs that London is not pretty and that London is pretty if New York is. Similarly here, if we pay attention only to Pierre's behaviour as a French speaker (and at least in his monolingual days he was no different from any other Frenchmen), Pierre satisfies all the normal criteria for believing that '*Londres*' has a referent uniquely satisfying the properties of being the largest city in England, containing Buckingham Palace, and the like. (If Pierre did not hold such beliefs, no Frenchman *ever* did.) Similarly, on the basis of his (later) beliefs expressed in English, Pierre also believes that the referent of 'London' uniquely satisfies these same properties. But Pierre cannot combine the two beliefs into a single set of beliefs from which he can draw the normal conclusion that 'London' and '*Londres*' must have the same referent. (Here the trouble comes not from 'London' and '*Londres*' but from 'England' and '*Angleterre*' and the rest.) Indeed, if he *did* draw what would appear to be the normal conclusion in this case and any of the other cases, Pierre would in fact be guilty of a logical fallacy.

Of course the description theorist could hope to eliminate the problem by 'defining' '*Angleterre*', 'England', and so on by appropriate descriptions also. Since in principle the problem may rear its head at the next 'level' and at each subsequent level, the description theorist would have to believe that an 'ultimate' level can eventually be reached where the defining properties are 'pure' properties not involving proper names (nor natural kind terms or related terms, see below!). I know of no convincing reason to suppose that such a level can be reached in any plausible way, or that the properties can continue to be uniquely identifying if one attempts to eliminate all names and related devices.[29] Such speculation aside, the fact remains that Pierre, judged by the *ordinary* criteria for such judgements, *did* learn both '*Londres*' and 'London' by *exactly* the same set of identifying properties; yet the puzzle remains even in this case.

Well, then, is there any way out of the puzzle? Aside from the principles of disquotation and translation, only our normal practice of translation of French into English has been used. Since the principles of disquotation and translation seem self-evident, we may be tempted to blame the trouble on the translation of '*Londres est jolie*' as 'London is pretty,' and ultimately, then, on the translation of '*Londres*' as 'London.'[30] Should we, perhaps, permit ourselves to conclude that '*Londres*' should not, 'strictly speaking' be translated as 'London'? Such an expedient is, of course, desperate: the translation in question is a standard one, learnt by students together with other standard translations of French into English. Indeed, '*Londres*' is, in effect, introduced into French as the French version of 'London'.

Since our backs, however, are against the wall, let us consider this desperate and implausible expedient a bit further. If '*Londres*' is *not* a correct French version of the English 'London,' under what circumstances can proper names be translated from one language to another?

Classical description theories suggest the answer: Translation, strictly speaking, is between idiolects; a name in one idiolect can be translated into another when (and only when) the speakers of the two idiolects associate the same uniquely identifying properties with the two names. We have seen that any such proposed restriction, not only fails blatantly to fit our normal practices of translation and indirect discourse reportage, but does not even appear to block the paradox.[31]

So we still want a suitable restriction. Let us drop the references to idiolects and return to '*Londres*' and 'London' as names in French and English, respectively—the languages of two communities. If '*Londres*' is not a correct French translation of 'London', could any other version do better? Suppose I introduced another word into French, with the stipulation that *it* should always be used to translate 'London'. Would not the same problem arise for this word as well? The only feasible solution in this direction is the most drastic: decree that no sentence containing a name can be translated except by a sentence containing the phonetically identical name. Thus when Pierre asserts '*Londres est jolie*', we English speakers can at best conclude, if anything: Pierre believes that *Londres* is pretty. Such a conclusion is, of course, not expressed in English, but in a word salad of English and French; on the view now being entertained, we cannot state Pierre's belief in *English* at all.[32] Similarly, we would have to say: Pierre believes that *Angleterre* is a monarchy, Pierre believes that *Platon* wrote dialogues, and the like.[33]

This 'solution' appears at first to be effective against the paradox, but it is drastic. What is it about sentences containing names that makes them—a substantial class—intrinsically untranslatable, express beliefs that cannot be reported in any other language? At best, to report them in the other language, one is forced to use a word salad in which names from the one language are imported into the other. Such a supposition is both contrary to our normal practice of translation and very implausible on its face.

Implausible though it is, there is at least this much excuse for the 'solution' at this point. Our normal practice with respect to some famous people and especially for geographical localities is to have different names for them in different languages, so that in translating sentences we translate the names. But for a large number of names, especially names of people, this is not so: the person's name is used in the sentences of all languages. At least the restriction in question merely urges us to mend our ways by doing *always* what we presently do *sometimes*.

But the really drastic character of the proposed restriction comes out when we see how far it may have to extend. In 'Naming and Necessity' I suggested that there are important analogies between proper names and natural kind terms, and it seems to me that the present puzzle is one instance where the analogy will hold. Putnam, who has proposed views on natural kinds similar to my own in many respects, stressed this extension of the puzzle in his comments at the Second Jerusalem Philosophical Encounter (April 1976). Not that the puzzle extends to all translation from English to French. At the moment, at least, it seems to me that Pierre, if he learns English and French separately, without learning any translation manual between them, *must* conclude, if he reflects enough, that 'doctor' and '*médecin*', and '*heureux*' and 'happy', are synonymous, or at any rate, coextensive;[34] any potential paradox of the present kind for these word pairs is thus blocked. But what about '*lapin*' and 'rabbit', or 'beech' and '*hêtre*'? We may suppose that Pierre is himself neither a zoologist nor a botanist. He has learnt each language in its own country, and the examples he has been shown to illustrate '*les lapins*' and 'rabbits,' 'beeches' and '*les hêtres*' are distinct. It thus seems to be possible for him to suppose that '*lapin*' and 'rabbit,' or 'beech' and '*hêtre*', denote distinct but superficially similar kinds or species, even though the differences may be indiscernible to the untrained eye. (This is especially plausible if, as Putnam supposes, an English speaker—for example, Putnam himself—who is not a botanist may use 'beech' and 'elm' with their normal (distinct) meanings, even though he cannot himself distinguish the two trees.[35] Pierre may quite plausibly be supposed to wonder whether the trees which in France he called '*les hêtres*' were beeches or elms, even though as a speaker of French he satisifies all usual criteria for using '*les hêtres*' normally. If beeches and elms will not serve, better pairs of ringers exist that cannot be told apart except by an expert.) Once Pierre is in such a situation, paradoxes analogous to the one about London obviously can arise for rabbits and beeches. Pierre could affirm a French statement with '*lapin*', but deny its English translation with 'rabbit.' As above, we are hard pressed to say what Pierre *believes*. We were considering a 'strict and philosophical' reform of translation procedures which proposed that foreign proper names should always be appropriated rather than translated. Now it seems that we will be forced to do the same with all words for natural kinds. (For example, on price of paradox, one must not translate '*lapin*' as 'rabbit'!) No longer can the extended proposal be defended, even weakly, as 'merely' universalizing what we already do sometimes. It is surely too drastic a change to retain any credibility.[36]

There is yet another consideration that makes the proposed restriction more implausible: Even this restriction does not really block the paradox. Even if we confine ourselves to a single language, say English, and to phonetically identical tokens of a single name, we can still generate the puzzle. Peter (as we may as well say now) may learn the name 'Paderewski' with an identification of the person named as a famous pianist. Naturally, having learnt this, Peter will assent to "Paderewski had musical talent", and *we* can infer—using 'Paderewski', as we usually do, to name the Polish musician and statesman:

(8) Peter believes that Paderewski had musical talent.

Only the disquotational principle is necessary for our inference; no translation is required. Later, in a different circle, Peter learns of someone called 'Paderewski' who was a Polish nationalist leader and Prime Minister. Peter is sceptical of the musical abilities of politicians. He concludes that probably two people, approximate contemporaries no doubt, were both named 'Paderewski'. Using 'Paderewski' as a name for the *statesman*, Peter assents to, "Paderewski had no musical talent". Should we infer, by the disquotational principle,

(9) Peter believes that Paderewski had no musical talent

or should we not? If Peter had not had the past history of learning the name 'Paderewski' in another way, we certainly would judge him to be using 'Paderewski' in a normal way, with the normal reference, and we would infer (9) by the disquotational principle. The situation is parallel to the problem with Pierre and London. Here, however, no restriction that names should not be translated, but should be phonetically repeated in the translation, can help us. Only a single language and a single name are involved. If any notion of translation is involved in this example, it is homophonic translation. Only the disquotational principle is used explicitly.[37] (On the other hand, the original 'two languages' case had the advantage that it would apply even if we spoke languages in which all names must denote uniquely and unambiguously.) The restriction that names must not be translated is thus ineffective, as well as implausible and drastic.

I close this section with some remarks on the relation of the present puzzle to Quine's doctrine of the 'indeterminacy of translation', with its attendant repudiation of intensional idioms of 'propositional attitude' such as belief and even indirect quotation. To a sympathizer with these doctrines the present puzzle may well seem to be just more grist for a familiar mill. The situation of the puzzle seems to lead to a breakdown of our normal practices of attributing belief and even of indirect quotation. No obvious paradox arises if we describe the same situation in terms of Pierre's sincere assent to various sentences, together with the conditions under which he has learnt the name in question. Such a description, although it does not yet conform to Quine's strict behaviouristic standards, fits in well with his view that in some sense direct quotation is a more 'objective' idiom than the propositional attitudes. Even those who, like the present writer, do not find Quine's negative attitude to the attitudes completely attractive must surely acknowledge this.

But although sympathizers with Quine's view can use the present examples to support it, the differences between these examples and the considerations Quine adduces for his own scepticism about belief and translation should not escape us. Here we make no use of hypothetical exotic systems of translation differing radically from the usual one, translating '*lapin*', say, as 'rabbit stage' or 'undetached part of a rabbit'. The problem arises entirely within our usual and customary system of translation of French into English; in one case, the puzzle arose even within English alone, using at most 'homophonic' translation. Nor is the problem that many different interpretations or translations fit our usual criteria, that, in Davidson's phrase,[38] there is more than one 'way of getting it right'. The trouble here is not that many views as to Pierre's beliefs get it right, but that they all definitely get it *wrong*. A straightforward application of the principles of translation and disquotation to all Pierre's utterances, French and English, yields the result that Pierre holds inconsistent beliefs, that logic alone should teach him that one of his beliefs is false. Intuitively, this is plainly incorrect. If we refuse to apply the principles to his French utterances at all, we would conclude that Pierre never believed that London is pretty, even though, before his unpredictable move, he was like any other monolingual Frenchman. This is absurd. If we refuse to ascribe the belief in London's pulchritude only after Pierre's move to England, we get the counterintuitive result that Pierre has changed his mind, and so on. But we have surveyed the possibilities above: the point was not that they are 'equally good', but that all are *obviously wrong*. If the puzzle is to be used as an argument for a Quinean position, it is an argument of a fundamentally different kind from those given before. And even Quine, if he wishes to incorporate the notion of belief even into a 'second level' of canonical notation,[39] must regard the puzzle as a real problem.

The alleged indeterminacy of translation and indirect quotation causes relatively little trouble for such a scheme for belief; the embarrassment it presents to such a scheme is, after all, one of riches. But the present puzzle indicates that the usual principles we use to ascribe beliefs are apt, in certain cases, to lead to contradiction, or at least, patent falsehoods. So it presents a problem for any project, Quinean or other, that wishes to deal with the 'logic' of belief on any level.[40]

IV. CONCLUSION

What morals can be drawn? The primary moral—quite independent of any of the discussion of the first two sections—is that the puzzle *is* a puzzle. As any theory of truth must deal with the Liar Paradox, so any theory of belief and names must deal with this puzzle.

But our theoretical starting point in the first two sections concerned proper names and belief. Let us return to Jones, who assents to "Cicero was bald" and to "Tully was not bald". Philosophers, using the disquotational principle, have concluded that Jones believes that Cicero was bald but that Tully was not. Hence, they have concluded, since Jones does not have contradictory beliefs, belief contexts are not 'Shakespearean' in Geach's sense: codesignative proper names are not interchangeable in these contexts *salva veritate*.[41]

I think the puzzle about Pierre shows that the simple conclusion was unwarranted. Jones's situation strikingly resembles Pierre's. A proposal that 'Cicero' and 'Tully' *are* interchangeable amounts roughly to a homophonic 'translation' of English into itself in which 'Cicero' is mapped into 'Tully' and *vice versa*, while the rest is left fixed. Such a 'translation' can, indeed, be used to obtain a paradox. But should the problem be blamed on this step? Ordinarily we would suppose without question that sentences in French with '*Londres*' should be translated into English with 'London'. Yet the same paradox results when we apply this translation too. We have seen that the problem can even arise with a single name in a single language, and that it arises with natural kind terms in two languages (or one: see below).

Intuitively, Jones's assent to both 'Cicero was bald' and 'Tully was not bald' arises from sources of just the same kind as Pierre's assent to both '*Londres est jolie*' and 'London is not pretty.'

It is wrong to blame unpalatable conclusions about Jones on substitutivity. The reason does not lie in any specific fallacy in the argument but rather in the nature of the realm being entered. Jones's case is just like Pierre's: both are in an area where our normal practices of attributing belief, based on the principles of disquotation and translation or on similar principles, are questionable.

It should be noted in this connection that the principles of disquotation and translation can lead to 'proofs' as well as 'disproofs' of substitutivity in belief contexts. In Hebrew there are two names for Germany, transliteratable roughly as '*Ashkenaz*' and '*Germaniah*'—the first of these may be somewhat archaic. When Hebrew sentences are translated into English, both become '*Germany*'. Plainly a normal Hebrew speaker analogous to Jones might assent to a Hebrew sentence involving '*Ashkenaz*' while dissenting from its counterpart with '*Germaniah*'. So far there is an argument *against* substitutivity. But there is also an argument *for* substitutivity, based on the principle of translation. Translate a Hebrew sentence involving '*Ashkenaz*' into English, so that '*Ashkenaz*' goes into 'Germany'. Then retranslate the result into Hebrew, this time translating 'Germany' as '*Germaniah*'. By the principle of translation, both translations preserve truth value. So: the truth value of any sentence of Hebrew involving '*Ashkenaz*' remains the same when '*Ashkenaz*' is replaced by '*Germaniah*'—a 'proof' of substitutivity! A similar 'proof' can be provided wherever there are two names in one language, and a normal practice of translating both indifferently into a single name of another language.[42] (If we combine the 'proof' and 'disproof' of substitutivity in this paragraph, we could get yet another paradox analogous to Pierre's: our Hebrew speaker both believes, and disbelieves, that Germany is pretty. Yet no amount of pure logic or semantic introspection suffices for him to discover his error.)

Another consideration, regarding natural kinds: Previously we pointed out that a bilingual may learn '*lapin*' and 'rabbit' normally in each respective language yet wonder whether they are one species or two, and that this fact can be used to generate a paradox analogous to Pierre's. Similarly, a speaker of *English* alone may learn 'furze' and 'gorse' normally (separately), yet wonder whether these are the same, or resembling kinds. (What about 'rabbit' and 'hare'?) It would be easy for such a speaker to assent to an assertion formulated with 'furze' but withhold assent from the corresponding assertion involving 'gorse'. The situation is quite analogous to that of Jones with respect to 'Cicero' and 'Tully'. Yet 'furze' and 'gorse', and other pairs of terms for the same natural kind, are normally thought of as *synonyms*.

The point is *not*, of course, that codesignative proper names *are* interchangeable in belief contexts *salva veritate*, or that they *are* interchangeable in simple contexts even *salva significatione*. The point is that the absurdities that disquotation plus substitutivity would generate are exactly paralleled by absurdities generated by disquotation plus translation, or even 'disquotation alone' (or: disquotation plus homophonic translation). Also, though our naïve practice may lead to 'disproofs' of substitutivity in certain cases, it can also lead to 'proofs' of substitutivity in some of these same cases, as we saw two paragraphs back. When we enter into the area exemplified by Jones and Pierre, we enter into an area where our normal practices of interpretation and attribution of belief are subjected to the greatest possible strain, perhaps to the point of breakdown. So is the notion of the *content* of someone's assertion, the *proposition* it expresses. In the present state of our knowledge, I think it would be foolish to draw any conclusion, positive or negative, about substitutivity.[43]

Of course nothing in these considerations prevents us from observing that Jones can sincerely assert both "Cicero is bald" and "Tully is not bald", even though he is a normal speaker of English and uses 'Cicero' and 'Tully' in normal ways, and with the normal referent. Pierre and the other paradoxical cases can be described similarly. (For those interested in one of my own doctrines, we can still say that there was a time when men were in no epistemic position to assent to 'Hesperus is Phosphorus' for want of empirical information, but it nevertheless expressed a necessary truth.)[44] But it is no surprise that quoted contexts fail to satisfy a substitutivity principle within the quotation marks. And, in our *present* state of clarity about the problem, we are in no position to apply a disquotation principle to these cases, nor to judge when two such sentences do, or do not, express the same 'proposition.'

Nothing in the discussion impugns the conventional judgment that belief contexts are 'referentially opaque,' if 'referential opacity' is construed so that failure of coreferential *definite descriptions* to be interchangeable *salva veritate* is sufficient for referential opacity. No doubt Jones can believe that the number of planets is even, without believing that the square of three is even, if he is under a misapprehension about the astronomical, but not the arithmetical facts. The question at hand was whether belief contexts were 'Shakespearean', not whether they were 'referentially transparent'. (Modal contexts, in my opinion, are 'Shakespearean' but 'referentially opaque'.)[45]

Even were we inclined to rule that belief contexts are not Shakespearean, it would be implausible at present to use the phenomenon to support a Frege–Russellian theory that names have descriptive 'senses' through 'uniquely identifying properties'. There are the well-known arguments against description theories, independent of the present discussion; there is the implausibility of the view that difference in names is difference in idiolect; and finally, there are the arguments of the present paper that differences of associated properties do not explain the problems in any case. Given these considerations, and the cloud our paradox places over the notion of 'content' in this area, the relation of substitutivity to the dispute between Millian and Fregean conclusions is not very clear.

We repeat our conclusions: Philosophers have often, basing themselves on Jones's and similar cases, supposed that it goes virtually without saying that belief contexts are not 'Shakespearean'. I think that, at present, such a definite conclusion in unwarranted. Rather Jones's case, like Pierre's, lies in an area where our normal apparatus for the ascription of belief is placed under the greatest strain and may even break down. There is even less warrant at the present time, in the absence of a better understanding of the paradoxes of this paper, for the use of alleged failures of substitutivity in belief contexts to draw any significant theoretical conclusion about proper names. Hard cases make bad law.[46]

Notes

1. 'Naming and Necessity', in D. Davidson and G. Harman (eds.) *The Semantics of Natural Languages* (Dordrecht: Reidel, 1971), 253–355 and 763–9. (Also as a separate monograph published by Harvard University Press and Basil Blackwell, 1972, 1980). 'Identity and Necessity', in M. Munitz (ed.), *Identity and Individuation* (New York University Press, 1971), 135–64. Acquaintance with these papers is not a prerequisite for understanding the central puzzle of the present paper, but is helpful for understanding the theoretical background.
2. Frege gives essentially this example as the second footnote of 'On Sense and Reference'. For the "Who is . . .?" to be applicable one must be careful to elicit from one's informant properties that he regards as defining the name and determining the referent, not mere well-known facts about the referent. (Of course this distinction may well seem fictitious, but it is central to the original Frege–Russell theory.)
3. For convenience Russell's terminology is assimilated to Frege's. Actually, regarding genuine or 'logically proper' names, Russell is a strict Millian: 'logically proper names' *simply* refer (to immediate objects of acquaintance). But, according to Russell, what are ordinarily called 'names' are not genuine, logically proper names but disguised definite descriptions. Since Russell also regards definite descriptions as in turn disguised notation, he does not associate any 'senses' with descriptions, since they are not genuine singular terms. When all disguised notation is eliminated, the only singular terms remaining are logically proper names, for which no notion of 'sense' is required. When we speak of Russell as assigning 'senses' to names, we mean ordinary names and for convenience we ignore his view that the descriptions abbreviating them ultimately disappear on analysis.

   On the other hand, the explicit doctrine that names are abbreviated definite descriptions is due to Russell. Michael Dummett, in his recent *Frege* (Duckworth and Harper and Row, 1973), 110–11, denies that Frege held a description theory of senses. Although as far as I know Frege indeed makes no explicit statement to that effect, his examples of names conform to the doctrine, as Dummett acknowledges. Especially his 'Aristotle' example is revealing. He defines 'Aristotle' just as Russell would; it seems clear that in the case of a famous historical figure, the 'name' is indeed to be given by answering, in a uniquely specifying way, the 'who is' question. Dummett himself characterizes a sense as a 'criterion . . . such that the referent of the name, if any, is whatever object satisfies that criterion'. Since presumably the satisfaction of the criterion must be unique (so a unique referent is determined), does not this amount to defining names by unique satisfaction of properties, *i.e.*, by descriptions? *Perhaps* the point is that the property in question need not be expressible by a usual predicate of English, as might be plausible if the referent is one of the speaker's acquaintances rather than a historical figure. But I doubt that even Russell, father of the explicitly formulated description theory, ever meant to require that the description must always be expressible in (unsupplemented) English.

In any event, the philosophical community has generally understood Fregean senses in terms of descriptions, and we deal with it under this usual understanding. For present purposes this is more important than detailed historical issues. Dummett acknowledges (p. 111) that few substantive points are affected by his (allegedly) broader interpretation of Frege; and it would not seem to be relevant to the problems of the present paper.

4. See Frege's footnote in 'On Sense and Reference' mentioned in n. 2, above and especially his discussion of 'Dr Gustav Lauben' in '*Der Gedanke*'. (In the recent Geach–Stoothoff translation, 'Thoughts', in *Logical Investigations* (Oxford: Blackwell; 1977), 11–12) also in this volume.
5. Russell, as a Millian with respect to genuine names, accepts this argument with respect to 'logically proper names'. For example, taking for the moment 'Cicero' and 'Tully' as 'logically proper names', Russell would hold that if I judge that Cicero admired Tully, I am related to Cicero, Tully, and the admiration relation in a certain way: since Cicero *is* Tully, I am related in exactly the same way to Tully, Cicero, and admiration; therefore I judge that Tully admired Cicero. Again, if Cicero *did* admire Tully, then according to Russell a single fact corresponds to all of 'Cicero admired Tully', 'Cicero admired Cicero', etc. Its constituent (in addition to admiration) is the man Cicero, taken, so to speak, twice.

   Russell thought that 'Cicero admired Tully' and 'Tully admired Cicero' are in fact obviously not interchangeable. For him, this was one argument that 'Cicero' and 'Tully' are *not* genuine names, and that the Roman orator is no constituent of propositions (or 'facts' or 'judgements') corresponding to sentences containing the name.
6. Given the arguments of Church and others, I do not believe that the formal mode of speech is synonymous with other formulations. But it can be used as a rough way to convey the idea of scope.
7. It may well be argued that the Millian view implies that proper names are *scopeless* and that for them the *de dicto–de re* distinction vanishes. This view has considerable plausibility (my own views on rigidity will imply something like this for *modal* contexts), but it need not be argued here either way: *de re* uses are simply not treated in the present paper.

   Christopher Peacocke ('Proper Names, Reference, and Rigid Designation', in S. Blackburn (ed.), *Meaning, Reference, and Necessity* (Cambridge, 1975); see Section I), uses what amounts to the equivalence of the *de dicto–de re* constructions in *all* contexts (or, put alternatively, the lack of such a distinction) to characterize the notion of rigid designation. I agree that for *modal* contexts, this is (roughly) equivalent to my own notion, also that for proper names Peacocke's equivalence holds for temporal contexts. (This is roughly equivalent to the 'temporal rigidity' of names.) I also agree that it is very plausible to extend the principle to all contexts. But, as Peacocke recognizes, this appears to imply a substitutivity principle for codesignative proper names in belief contexts, which is widely assumed to be false. Peacocke proposes to use Davidson's theory of intensional contexts to block this conclusion (the material in the 'that' clause is a separate sentence). I myself cannot accept Davidson's theory; but even if it were true, Peacocke in effect acknowledges that it does not really dispose of the difficulty (p. 127, first paragraph). (Incidentally, if Davidson's theory does block any inference to the transparency of belief contexts with respect to names, why does Peacocke assume without argument that it does not do so for modal contexts, which have a similar grammatical structure?) The problems are thus those of the present paper; until they are resolved I prefer at present to keep to my earlier more cautious formulation.

Incidentally, Peacocke hints a recognition that the received platitude—that codesignative names are not interchangeable in belief contexts—may not be so clear as is generally supposed.

8. The example comes from Quine, *Word and Object* (MIT Press, 1960), 145. Quine's conclusion that 'believes that' construed *de dicto* is opaque has widely been taken for granted. In the formulation in the text I have used the colon to emphasize that I am speaking of belief *de dicto*. Since, as I have said, belief *de dicto* will be our *only* concern in this paper, in the future the colon will usually be suppressed, and all 'believes that' contexts should be read *de dicto* unless the contrary is indicated explicitly.
9. In many writings Peter Geach has advocated a view that is non-Millian (he would say 'non-Lockean') in that to each name a sortal predicate is attached by definition ('Geach', for example, by *definition* names a man). On the other hand, the theory is not completely Fregean either, since Geach denies that any definite description that would identify the referent of the name among things of the same sort is analytically tied to the name. (See e.g. his *Reference and Generality* (Cornell, 1962), 43–5.) As far as the present issues are concerned, Geach's view can fairly be assimilated to *Mill*'s rather than Frege's. For such ordinary names as 'Cicero' and 'Tully' will have both the same reference and the same (Geachian) sense (namely, that they are names of a man). It would thus seem that they ought to be interchangeable everywhere. (In *Reference and Generality*, Geach appears not to accept this conclusion, but the *prima facie* argument for the conclusion will be the same as on a purely Millian view.)
10. In an unpublished paper, Diana Ackerman urges the problem of substitutivity failure against the Millian view and, hence, against my own views. I believe that others may have done so as well. (I have the impression that the paper has undergone considerable revision, and I have not seen recent versions.) I agree that this problem is a considerable difficulty for the Millian view, and for the Millian *spirit* of my own views in 'Naming and Necessity'. (See the discussion of this in the text of the present paper.) On the other hand I would emphasize that there need be no *contradiction* in maintaining that names are *modally* rigid and satisfy a substitutivity principle for modal contexts, while denying the

substitutivity principle for belief contexts. The entire apparatus elaborated in 'Naming and Necessity' of the distinction between epistemic and metaphysical necessity, and of giving a meaning and fixing a reference, was meant to show, among other things, that a Millian substitutivity doctrine for modal contexts can be maintained even if such a doctrine for epistemic contexts is rejected. 'Naming and Necessity' never asserted a substitutivity principle for epistemic contexts.

It is even consistent to suppose that differing modes of (rigidly) fixing the reference is responsible for the substitutivity failures, thus adopting a position intermediate between Frege and Mill, on the lines indicated in the text of the present paper. 'Naming and Necessity' may even perhaps be taken as suggesting, for some contexts where a conventional description rigidly fixes the reference ('Hesperus–Phosphorus'), that the mode of reference fixing is relevant to epistemic questions. I knew when I wrote 'Naming and Necessity' that substitutivity issues in epistemic contexts were really very delicate, due to the problems of the present paper, but I thought it best not to muddy the waters further. (See nn. 43–4.)

After this paper was completed, I saw Alvin Plantinga's paper 'The Boethian Compromise', *American Philosophical Quarterly* 15 (1978), 129–38. Plantinga adopts a view intermediate between Mill and Frege, and cites substitutivity failures as a principal argument for his position. He also refers to a forthcoming paper by Ackerman. I have not seen this paper, but it probably is a descendant of the paper referred to above.

11. Here I use 'connotation' so as to imply that the associated properties have an *a priori* tie to the name, at least as rigid reference fixers, and therefore must be true of the referent (if it exists). There is another sense of 'connotation,' as in 'The Holy Roman Empire', where the connotation need not be assumed or even believed to be true of the referent. In some sense akin to this, classicists and others with some classical learning may attach certain distinct 'connotations' to 'Cicero' and 'Tully'. Similarly, 'The Netherlands' may suggest low altitude to a thoughtful ear. Such 'connotations' can hardly be thought of as community-wide; many use the names unaware of such suggestions. Even a speaker aware of the suggestion of the name may not regard the suggested properties as true of the object; cf. 'The Holy Roman Empire.' A 'connotation' of this type neither gives a meaning nor fixes a reference.

12. Some might attempt to find a difference in 'sense' between 'Cicero' and 'Tully' on the grounds that "Cicero is called 'Cicero' " is trivial, but "Tully is called 'Cicero' " may not be. Kneale, and in one place (probably at least implicitly) Church, have argued in this vein. (For Kneale, see 'Naming and Necessity', p. 283.) So, it may be argued, being called 'Cicero' is part of the sense of the name 'Cicero', but not part of that of 'Tully'.

    I have discussed some issues related to this in 'Naming and Necessity', pp. 283–6. (See also the discussions of circularity conditions elsewhere in 'Naming and Necessity'.) Much more could be said about and against this kind of argument; perhaps I will sometime do so elsewhere. Let me mention very briefly the following parallel situation (which may be best understood by reference to the discussion in 'Naming and Necessity'). Anyone who understands the meaning of 'is called' and of quotation in English (and that 'alienists' is meaningful and grammatically appropriate), knows that "alienists are called 'alienists' " expresses a truth in English, even if he has no idea what 'alienists' means. He need *not* know that "psychiatrists are called 'alienists'" expresses a truth. None of this goes to show that 'alienists' and 'psychiatrists' are not synonymous, or that 'alienists' has *being called 'alienists'* as part of its meaning when 'psychiatrists' does not. Similarly for 'Cicero' and 'Tully'. There is no more reason to suppose that being so-called is part of the meaning of a name than of any other word.

13. A view follows Frege and Russell on this issue, even if it allows each speaker to associate a cluster of descriptions with each name, provided that it holds that the cluster varies from speaker to speaker and that variations in the cluster are variations in idiolect. Searle's view thus is Frege–Russellian when he writes in the concluding paragraph of 'Proper Names' (*Mind* 67 (1958), 166–73), "'Tully = Cicero' would, I suggest, be analytic for most people; the same descriptive presuppositions are associated with each name. But of course if the descriptive presuppositions were different it might be used to make a synthetic statement."
14. Though here I use the jargon of propositions, the point is fairly insensitive to differences in theoretical standpoints. For example, on Davidson's analysis, I would be asserting (roughly) that many are unaware-of-the-content-of the following *utterance* of mine: Cicero is Tully. This would be subject to the same problem.
15. Benson Mates, 'Synonymity', *University of California Publications in Philosophy* 25 (1950), 201–26; reprinted in *Semantics and the Philosophy of Language*, L. Linsky (ed.) (University of Illinois Press, 1952). (There was a good deal of subsequent discussion. In Mates's original paper the point is made almost parenthetically.) Actually, I think that Mates's problem has relatively little force against the argument we are considering for the Fregean position. Mates's puzzle in no way militates against some such principle as: If one word is synonymous with another, then a sufficiently reflective speaker subject to no linguistic inadequacies or conceptual confusions who sincerely assents to a simple sentence containing the one will also (sincerely) assent to the corresponding sentence with the other in its place.

    It is surely a crucial part of the present 'Fregean' argument that codesignative names may have distinct 'senses', that a speaker may assent to a simple sentence containing one and deny the corresponding sentence containing the other, even though he is *guilty of no conceptual or linguistic confusion, and of no lapse in logical consistency*. In the case of two straightforward synonyms, this is not so.

    I myself think that Mates's argument is of considerable interest, but that the issues are confusing and delicate and that, if the argument works, it probably leads to a paradox or puzzle rather than to a definite conclusion. (See also nn. 23, 28, and 46.)
16. 'Naming and Necessity', pp. 291–3.
17. Recall also n. 12.
18. Some philosophers stress that names are not *words* of a language, or that names are not *translated* from one language to another. (The phrase 'common currency of our common language' was meant to be neutral with respect to any such alleged issue.) Someone may use 'Mao Tse-Tung', for example, in English, though he knows not one word of Chinese. It seems hard to deny, however, that "*Deutschland*", "*Allemagne*", and "Germany", are the German, French, and English names of a single country, and that one translates a French

sentence using "*Londres*" by an English sentence using "London". Learning these facts *is* part of learning German, French, and English.

It would appear that *some* names, especially names of countries, other famous localities, and some famous people, *are* thought of as part of a language (whether they are called 'words' or not is of little importance). Many other names are not thought of as part of a language, especially if the referent is not famous (so the notation used is confined to a limited circle), or if the same name is used by speakers of all languages. As far as I can see, it makes little or no *semantic* difference whether a particular name is thought of as part of a language or not. Mathematical notation such as ' < ' is also ordinarily not thought of as part of English, or any other language, though it is used in combination with English words in sentences of mathematical treatises written in English. (A French mathematician can use the notation though he knows not one word of English.) 'Is less than', on the other hand, *is* English. Does this difference have any semantic significance?

I will speak in most of the text as if the names I deal with are part of English, French, etc. But it matters little for what I say whether they are thought of as parts of the language or as adjuncts to it. And one need not say that a name such as '*Londres*' is 'translated' (if such a terminology suggested that names have 'senses', I too would find it objectionable), as long as one acknowledges that *sentences* containing it are properly translated into English using 'London'.

19. By saying that names are transparent in a context, I mean that codesignative names are interchangeable there. This is a deviation for brevity from the usual terminology, according to which the *context* is transparent. (I use the usual terminology in the paper also.)
20. But we must use the term 'sense' here in the sense of 'that which fixes the reference', not 'that which gives the meaning', otherwise we shall run afoul of the rigidity of proper names. If the source of a chain for a certain name is in fact a given object, we use the name to designate that object even when speaking of counterfactual situtions in which some *other* object originated the chain.
21. The point is that, according to the doctrine of 'Naming and Necessity', when proper names are transmitted from link to link, even though the beliefs about the referent associated with the name change radically, the change is not to be considered a linguistic change in the way it *was* a linguistic change when 'villain' changed its meaning from 'rustic' to 'wicked man'. As long as the reference of a name remains the same, the associated beliefs about the object may undergo a large number of changes without these changes constituting a change in the language.

    If Geach is right, an appropriate sortal must be passed on also. But see footnote 58 of 'Naming and Necessity'.
22. Similar appropriate restrictions are assumed below for the strengthened disquotational principle and for the principle of translation. Ambiguities need not be excluded if it is tacitly assumed that the sentence is to be understood in one way in all its occurrences. (For the principle of translation it is similarly assumed that the translator matches the *intended* interpretation of the sentence.) I do not work out the restrictions on indexicals in detail, since the intent is clear.

    Clearly, the disquotational principle applies only to *de dicto*, not *de re*, attributions of belief. If someone sincerely assents to the near triviality, "The tallest foreign spy is a spy", it follows that he believes that: the tallest foreign spy is a spy. It is well known that it does *not* follow that he believes, *of* the tallest foreign spy, that he is a spy. In the latter case, but not in the former, it would be his patriotic duty to make contact with the authorities.

23. What if a speaker assents to a sentence, but fails to assent to a synonymous assertion? Say, he assents to "Jones is a doctor", but not to "Jones is a physician". Such a speaker either does not understand one of the sentences normally, or he should be able to correct himself "on reflection". As long as he confusedly assents to 'Jones is a doctor' but not to 'Jones is a physician', we *cannot* straightforwardly apply disquotational principles to conclude that he does or does not believe that Jones is a doctor, because his assent is not "reflective".

    Similarly, if someone asserts, "Jones is a doctor but not a physician", he should be able to recognize his inconsistency without further information. We have formulated the disquotational principles so they need not lead us to attribute belief as long as we have grounds to suspect conceptual or linguistic confusion, as in the cases just mentioned..

    Note that if someone says, "Cicero was bald but Tully was not", there need be *no* grounds to suppose that he is under *any* linguistic or conceptual confusion.

24. This should not be confused with the question whether the speaker simultaneously believes *of* a given object, both that it has a certain property and that it does not have it. Our discussion concerns *de dicto* (notional) belief, not *de re* belief.

    I have been shown a passage in Aristotle that appears to suggest that *no one* can really believe both of two explicit contradictories. If we wish to use the *simple* disquotational principle as a test for disbelief, it suffices that this be true of *some* individuals, after reflection, who are simultaneously aware of both beliefs, and have sufficient logical acumen and respect for logic. Such individuals, if they have contradictory beliefs, will be shaken in one or both beliefs after they note the contradiction. For such individuals, sincere reflective assent to the negation of a sentence implies disbelief in the proposition it expresses, so the test in the text applies.

25. For example, in translating a historical report into another language, such as "Patrick Henry said, 'Give me liberty or give me death!' ", the translator may well translate the quoted material attributed to Henry. He translates a presumed truth into a falsehood, since Henry spoke English; but probably his reader is aware of this and is more interested in the content of Henry's utterance than in its exact words. Especially in translating fiction, where truth is irrelevant, this procedure is appropriate. But some objectors to Church's 'translation argument' have allowed themselves to be misled by the practice.

26. To state the argument precisely, we need in addition a form of the Tarskian disquotation principle for truth: For each (French or English) replacement for '*p*', infer " '*p*' is true" from "*p*", and conversely. (Note that " '*p*' is true" becomes an English sentence even if '*p*' is replaced by a French sentence.) In the text we leave the application of the Tarskian disquotational principle tacit.

27. I gather that Burali-Forti originally thought he had 'proved' that the ordinals are not linearly ordered, reasoning in a manner similar to our topologist. Someone who heard the present paper delivered told me that König made a similar error.

28. It is not possible, in this case, as it is in the case of the man who assents to

"Jones is a doctor" but not to "Jones is a physician", to refuse to apply the disquotational principle on the grounds that the subject must lack proper command of the language or be subject to some linguistic or conceptual confusion. As long as Pierre is unaware that 'London' and '*Londres*' are codesignative, he need not lack appropriate linguistic knowledge, nor need he be subject to any linguistic or conceptual confusion when he affirms '*Londres est jolie*' but denies 'London is pretty'.

29. The 'elimination' would be most plausible if we believed, according to a Russellian epistemology, that all my language, when written in unabbreviated notation, refers to constituents with which I am 'acquainted' in Russell's sense. Then no one speaks a language intelligible to anyone else; indeed, no one speaks the same language twice. Few today will accept this.

    A basic consideration should be stressed here. Moderate Fregeans atempt to combine a roughly Fregean view with the view that names are part of our common language, and that our conventional practices of interlinguistic translation and interpretation are correct. The problems of the present paper indicate that it is very difficult to obtain a requisite socialized notion of sense that will enable such a program to succeed. Extreme Fregeans (such as Frege and Russell) believe that in general names are peculiar to idiolects. They therefore would accept no general rule translating '*Londres*' as 'London,' nor even translating one person's use of 'London' into another's. However, if they follow Frege in regarding senses as 'objective', they must believe that in principle it makes sense to speak of two people using two names in their respective idiolects with the same sense, and that there must be (necessary and) sufficient conditions for this to be the case. If these conditions for sameness of sense are satisified, translation of one name into the other is legitimate, otherwise not. The present considerations (and the extension of these below to natural kind and related terms), however, indicate that the notion of sameness of sense, if it is to be explicated in terms of sameness of identifying properties and if these properties are themselves expressed in the languages of the two respective idiolects, presents interpretation problems of the same type presented by the names themselves. Unless the Fregean can give a method for identifying sameness of sense that is free of such problems, he *has no sufficient conditions for sameness of sense, nor for translation to be legitmate*. He would therefore be forced to maintain, contrary to Frege's intent, that not only in practice do few people use proper names with the same sense but that *it is principle meaningless to compare senses*. A view that the identifying properties used to define senses should always be expressible in a Russellian language of 'logically proper names' would be one solution to this difficulty but involves a doubtful philosophy of language and epistemology.

30. If any reader finds the term 'translation' objectionable with respect to names, let him be reminded that all I mean is that French sentences containing '*Londres*' are uniformly translated into English with 'London'.

31. The paradox would be blocked if we required that they define the names by the same properties expressed in the same words. There is nothing in the motivation of the classical description theories that would justify this extra clause. In the present case of French and English, such a restriction would amount to a decree that neither '*Londres*', nor any other conceivable French name, could be translated as 'London'. I deal with this view immediately below.

32. Word salads of two languages (like ungrammatical 'semisentences' of a single language) need not be unintelligible, though they are makeshifts with no fixed syntax. "If God did not exist, Voltaire said, *il faudrait l'inventer*." The meaning is clear.
33. Had we said, "Pierre believes that the country he calls '*Angleterre*' is a monarchy", the sentence would be English, since the French word would be mentioned but not used. But for this very reason we would not have captured the sense of the French original.
34. Under the influence of Quine's *Word and Object*, some may argue that such conclusions are not inevitable: perhaps he will translate '*médecin*' as 'doctor stage', 'undetached part of a doctor'! If a Quinean sceptic makes an empirical prediction that such reactions from bilinguals as a matter of fact can occur, I doubt that he will be proved correct. (I do not know what Quine would think. but see *Word and Object*, p. 74, first paragraph.) On the other hand, if the translation of 'médecin' as 'doctor' rather than 'doctor part' in this situation *is*, empirically speaking, inevitable, then even the advocate of Quine's thesis will have to admit that there is something special about one particular translation. The issue is not crucial to our present concerns, so I leave it with these sketchy remarks. But see also n. 36.
35. Putnam gives the example of elms and beeches in "The Meaning of 'Meaning' " (in *Language, Mind, and Knowledge* (University of Minnesota Press, 1975) also reprinted in Putnam's *Philosphical Papers*, ii (Cambridge University Press, 1975). See also Putnam's discussion of other examples on pp. 139–43; also my own remarks on 'fool's gold', tigers, etc. in 'Naming and Necessity', pp. 316–23.
36. It is unclear to me how far this can go. Suppose Pierre hears English spoken only in England, French in France, and learns both by direct method. (Suppose also that no one else in each country speaks the language of the other.) Must he be sure that 'hot' and '*chaud*' are coextensive? In practice he certainly would. But suppose somehow his experience is consistent with the following bizarre—and of course, false!—hypothesis: England and France differ atmospherically so that human bodies are affected very differently by their interaction with the surrounding atmosphere. (This would be more plausible if France were on another planet.) In particular, within reasonable limits, things that feel cold in one of the countries feel hot in the other, and *vice versa*. Things don't change their *temperature* when moved from England to France, they just *feel* different because of their effects on human physiology. Then '*chaud*', in French, would be true of the things that are called 'cold' in English! (Of course the present discussion is, for lack of space, terribly compressed. See also the discussion of 'heat' in 'Naming and Necessity'. We are simply creating, for the physical property 'heat', a situation analogous to the situation for natural kinds in the text.)

    If Pierre's experiences were arranged somehow so as to be consistent with the bizarre hypothesis, and he somehow came to believe it, he might simultaneously assent to '*C'est chaud*' and 'This is cold' without contradiction, even though he speaks French and English normally in each country separately.

    This case needs much more development to see if it can be set up in detail, but I cannot consider it further here. Was I right in assuming in the text that the difficulty could not arise for '*médecin*' and 'doctor'?

37. One might argue that Peter and we do speak different dialects, since in Peter's idiolect 'Paderewski' is used ambiguously as a name for a musician and a statesman (even though these are in fact the same), while in our language it us used unambiguously for a musician-statesman. The problem then would be whether Peter's dialect can be translated homophonically into our own. Before he hears of 'Paderewski-the-statesman', it would appear that the answer is affirmative for his (then unambiguous) use of 'Paderewski', since he did not differ from anyone who happens to have heard of Paderewski's musical achievements but not of his statesmanship. Similarly for his later use of 'Paderewski', if we ignore his earlier use. The problem is like Pierre's, and is essentially the same whether we describe it in terms of whether Peter satisfies the condition for the disquotational principle to be applicable, or whether homophonic translation of his dialect into our own is legitimate.
38. D. Davidson, 'On Saying That', in D. Davidson and J. Hintikka (eds.), *Words and Objections* Dordrecht: Reidel, 1969), 166.
39. In *Word and Object*, p. 221, Quine advocates a second level of canonical notation, "to dissolve verbal perplexities or facilitate logical deductions", admitting the propositional attitudes, even though he thinks them "baseless" idioms that should be excluded from a notation "limning the true and ultimate structure of reality".
40. In one respect the considerations mentioned above on natural kinds show that Quine's translation apparatus is insufficiently sceptical. Quine is sure that the native's *sentence* "Gavagai!" should be translated "Lo, a rabbit!", provided that its affirmative and negative stimulus meanings for the native match those of the English sentence for the Englishman; scepticism sets in only when the linguist proposes to translate the *general term* 'gavagai' as 'rabbit' rather than 'rabbit stage', 'rabbit part', and the like. But there is another possibility that is independent of (and less bizarre than) such sceptical alternatives. In the geographical area inhabited by the natives, there may be a species indistinguishable to the nonzoologist from rabbits but forming a distinct species. Then the 'stimulus meanings', in Quine's sense, of 'Lo, a rabbit!' and 'Gavagai!' may well be identical (to nonzoologists), especially if the ocular irradiations in question do not include a specification of the geographical locality. ('Gavagais' produce the same ocular irradiation patterns as rabbits.) Yet 'Gavagai!' and 'Lo, a rabbit!' are hardly synonymous; on typical occasions they will have opposite truth values.

    I believe that the considerations about names, let alone natural kinds, emphasized in 'Naming and Necessity' go against any simple attempt to base interpretation solely on maximizing agreement with the affirmations attributed to the native, matching of stimulus meanings, etc. The 'Principle of Charity' on which such methodologies are based was first enunciated by Neil Wilson in the special case of proper names as a formulation of the cluster-of-descriptions theory. The argument of 'Naming and Necessity' is thus directed against the simple 'Principle of Charity' for that case.
41. Geach introduced the term 'Shakespearean' after the line, "A rose / By any other *name*, would smell as sweet".

    Quine seems to define 'referentially transparent' contexts so as to imply that coreferential names and definite descriptions must be interchangeable *salva veritate*. Geach stresses that a context may be 'Shakespearean' but not 'referentially transparent' in this sense.

42. Generally such cases may be slightly less watertight than the 'London'–'*Londres*' case. '*Londres*' just is the French version of 'London', while one cannot quite say that the same relation holds between '*Ashkenaz*' and '*Germaniah*'. Nevertheless:

    (*a*) Our standard practice in such cases is to translate both names of the first language into the single name of the second.

    (*b*) Often no nuances of 'meaning' are discernible differentiating such names as '*Ashkenaz*' and '*Germaniah*', such that we would not say either that Hebrew would have been impoverished had it lacked one of them (or that English is impoverished because it has only one name for Germany), any more than a language is impoverished if it has only one word corresponding to 'doctor' and 'physician'. Given this, it seems hard to condemn our practice of translating both names as 'Germany' as 'loose'; in fact, it would seem that Hebrew just has two names for the same country where English gets by with one.

    (*c*) Any inclinations to avoid problems by declaring, say, the translation of '*Ashkenaz*' as 'Germany' to be loose should be considerably tempered by the discussion of analogous problems in the text.

43. In spite of this official view, perhaps I will be more assertive elsewhere.
    In the case of 'Hesperus' and 'Phosphorus' (in contrast to 'Cicero' and 'Tully'), where there is a case for the existence of conventional community-wide 'senses' differentiating the two—at least, two distinct modes of 'fixing the reference of two rigid designators'—it is more plausible to suppose that the two names are definitely not interchangeable in belief contexts. According to such a supposition, a belief that Hesperus is a planet is a belief that a certain heavenly body, rigidly picked out as seen in the evening in the appropriate season, is a planet; and similarly for Phosphorus. One may argue that translation problems like Pierre's will be blocked in this case, that '*Vesper*' must be translated as 'Hesperus,' not as 'Phosphorus'. As against this, however, two things:

    (*a*) We should remember that sameness of properties used to fix the reference does *not* appear to guarantee in general that paradoxes will not rise. So one may be reluctant to adopt a solution in terms of reference-fixing properties for this case if it does not get to the heart of the general problem.

    (*b*) The main issue seems to me here to be—how essential is a particular mode of fixing the reference to a correct learning of the name? If a parent, aware of the familiar identity, takes a child into the fields in the morning and says (pointing to the morning star), "That is called 'Hesperus' ", has the parent mistaught the language? (A parent who says, "Creatures with kidneys are called 'cordates', definitely has mistaught the language, even though the statement is extensionally correct.) To the extent that it is *not* crucial for correct language learning that a particular mode of fixing the reference be used, to that extent there is no 'mode of presentation' differentiating the 'content' of a belief about 'Hesperus' from one about 'Phosphorus'. I am doubtful that the original method of fixing the reference *must* be preserved in transmission of the name.

If the mode of reference fixing *is* crucial, it can be maintained that otherwise identical beliefs expressed with 'Hesperus' and with 'Phosphorus' have definite differences of 'content', at least in an epistemic sense. The conventional ruling against substitutivity could thus be maintained without qualms for some cases, though not as obviously for others, such as 'Cicero' and 'Tully'. But it is unclear to me whether even 'Hesperus' and 'Phosphorus' do have such conventional 'modes of presentation'. I need not take a definite stand, and the verdict may be different for different particular pairs of names. For a brief related discussion, see 'Naming and Necessity', p. 331, first paragraph.

44. However, some earlier formulations expressed disquotationally such as "It was once unknown that Hesperus is Phosphorus" are questionable in the light of the present paper (but see the previous note for this case). I was aware of this question by the time 'Naming and Necessity' was written, but I did not wish to muddy the waters further than necessary at that time. I regarded the distinction between epistemic and metaphysical necessity as valid in any case and adequate for the distinctions I wished to make. The considerations in this paper are relevant to the earlier discussion of the 'contingent *a priori*' as well; perhaps I will discuss this elsewhere.

45. According to Russell, definite descriptions are not genuine singular terms. He thus would have regarded any concept of 'referential opacity' that includes definite descriptions as profoundly misleading. He also maintained a substitutivity principle for 'logically proper names' in belief and other attitudinal contexts, so that for him belief contexts were as 'transparent', in any philosophically decent sense, as truth-functional contexts.

    Independently of Russell's views, there is much to be said for the opinion that the question whether a context is 'Shakespearean' is more important philosophically—even for many purposes for which Quine invokes his own concept—than whether it is 'referentially opaque'.

46. I will make some brief remarks about the relation of Benson Mates's problem (see n. 15) to the present one. Mates argued that such a sentence as (*) 'Some doubt that all who believe that doctors are happy believe that physicians are happy', may be true, even though 'doctors' and 'phyicians' are synonymous, and even though it would have been false had 'physicians' been replaced in it by a second occurrence of 'doctors'. Church countered that (*) could not be true, since its translation into a language with only one word for doctors (which would translate both 'doctors' and 'physicians') would be false. If *both* Mates's and Church's intuitions were correct, we might get a paradox analogous to Pierre's.

    Applying the principles of translation and disquotation to Mates's puzzle, however, involves many more complications than our present problem. First, if someone assents to 'Doctors are happy', but refuses assent to 'Physicians are happy', *prima facie* disquotation does not apply to him since he is under a linguistic or conceptual confusion. (See n. 23.) So there are as yet no grounds, merely because this happened, to doubt that all who believe that doctors are happy believe that physicians are happy.

    Now suppose someone assents to 'Not all who believe that doctors are happy believe that physicians are happy'. What is the source of his assent? If it is failure to realize that 'doctors' and 'physicians' are synonymous (this was the situation Mates originally envisaged), then he is under a linguistic or conceptual confusion, so disquotation does not clearly apply. Hence we have no reason to conclude from this case that (*) is true. Alternatively, he may realize that 'doctors' and 'physicians' are synonymous; but he applies disquotation to a man who assents to 'Doctors are happy' but not to 'Physicians are happy', ignoring the caution of the previous paragraph. Here he is not under a simple linguistic confusion (such as failure to realize that 'doctors' and 'physicians' are synonymous), but he appears to be under a deep conceptual confusion (misapplication of the disquotational principle). Perhaps, it may be argued, he misunderstands the 'logic of belief'. Does his conceptual confusion mean that we cannot straightforwardly apply disquotation to his utterance, and that therefore we cannot conclude from his behaviour that (*) is true? I think that, although the issues are delicate, and I am not at present completely sure what answers to give, there is a case for an affirmative answer. (Compare the more extreme case of someone who is so confused that he thinks that someone's *dissent* from 'Doctors are happy' implies that he believes that doctors are happy. If someone's utterance, 'Many believe that doctors are happy', is based on such a misapplication of disquotation, surely we in turn should not apply disquotation to it. The utterer, at least in this context, does not really know what 'belief' means.)

I do *not* believe the discussion above ends the matter. Perhaps I can discuss Mates's problem at greater length elsewhere. Mates's problem is perplexing, and its relation to the present puzzle is interesting. But it should be clear from the preceding that Mates's argument involves issues even more delicate than those that arise with respect to Pierre. First, Mates's problem involves delicate issues regarding iteration of belief contexts, whereas the puzzle about Pierre involves the application of disquotation only to affirmations of (or assents to) *simple* sentences. More important, Mates's problem would not arise in a world where no one ever was under a linguistic or a conceptual confusion, no one ever thought anyone else was under such a confusion, no one ever thought anyone ever thought anyone was under such a confusion, and so on. It is important, both for the puzzle about Pierre and for the Fregean argument that 'Cicero' and 'Tully' differ in 'sense', that they would still arise in such a world. They are entirely free of the delicate problem of applying disquotation to utterances directly or indirectly based on the existence of linguistic confusion. See nn. 15 and 28, and the discussion in the text of Pierre's logical consistency.

Another problem discussed in the literature to which the present considerations may be relevant is that of 'self-consciousness', or the peculiarity of 'I'. Discussions of this problem have emphasized that 'I', even when Mary Smith uses it, is not interchangeable with 'Mary Smith', nor with any other conventional singular term designating Mary Smith. If she is 'not aware that she is Mary Smith', she may assent to a sentence with 'I', but dissent from the corresponding sentence with 'Mary Smith'. It is quite possible that any attempt to clear up the logic of all this will involve itself in the problem of the present paper. (For this purpose, the present discussion might be extended to demonstratives and indexicals.)

The writing of this paper had partial support from a grant from the National Science Foundation, a John Simon Guggenheim Foundation Fellowship, a Visiting Fellowship at All Souls College, Oxford, and a sabbatical leave from Princeton University. Various people at the Jerusalem Encounter and elsewhere, who will not be enumerated, influenced the paper through discussion.

VIII

SYNONYMY AND THE ANALYSIS OF BELIEF SENTENCES

HILARY PUTNAM

IN the paper 'Carnap's Analysis of Statements of Assertion and Belief',[1] Church has advanced some criticisms of the theory of belief sentences and indirect discourse proposed by Carnap.[2] It appears that these criticisms can be met without a modification of the theory. But certain criticisms by Benson Mates[3] would seem to be more serious; these criticisms are equally forceful against the widely held view that expressions with the same sense are interchangeable in all contexts. In this paper, a revision of this principle will be suggested; and a new definition of intensional isomorphism[4] will be put forward as a basis for the reconstruction of the theory of meaning-analysis in accordance with the suggested principle.

I

Church's paper on Carnap's analysis is divided, conveniently for our purposes, into two parts. The criticisms in part I do not apply to Carnap's conception, since Carnap intends that, for the purposes of reconstruction, a "language" is to be thought of as having precise formation rules, designation rules, and truth-rules, or as having had these made precise; and it is to be defined by reference to its rules rather than as "the language spoken in the British Isles in 1941", or something of that kind. The criticism in part II of Church's paper, however, is supposed to apply to Carnap's analysis even when "English" and "German" are

Hilary Putnam, 'Synomymy and the Analysis of Belief Sentences', *Analysis* 14 (1954), 114–22. (Original title: 'Synonymity, and the Analysis of Belief Sentences'.)

construed as semantical systems. Let us therefore turn to this part of Church's criticism.

Church begins by considering the sentence in the system E:[5]

(1) Seneca said that man is a rational animal

and its counterpart in the system G:

(1′) Seneca hat gesagt dass der Mensch ein vernuenftiges Tier sei.

The analysis of (1) in the system E by the technique proposed in Carnap, *Meaning and Necessity*, leads to the following sentence (we follow Church's enumeration):

(7) There is a sentence S_i in a semantical system S such that (*a*) S_i as sentence of S is intensionally isomorphic to 'Man is a rational animal' as sentence of E, and (*b*) Seneca wrote S_i as sentence of S

and similarly, the analysis of (1′) in the system G leads to:

(7′) Es gibt einen Satz S_i in einem semantischen System S, so dass (*a*) S_i als Satz von S intensional isomorph zu 'Der Mensch ist ein vernuenftiges Tier' als Satz in G ist, und (*b*) Seneca S_i als Satz von S geschrieben hat.

But (7) and (7′), as Church remarks, are "not intensionally isomorphic".[6]

But why should they be? Suppose someone proposes the following as an analysis of 'one' in the simplified theory of types (with "Systematic Ambiguity"):

(*a*) $\hat{x}\ (\exists y)\ (z)\ (z \epsilon x \equiv z = y)$

and suppose further that someone else proposes instead:

(*b*) $\hat{x}\ (\exists y)\ (y \epsilon x .\ (z)\ (z \epsilon x \supset z = y))$

These are, of course, "not intensionally isomorphic". Yet there would be no contradiction involved in regarding both analyses as correct, for they are logically equivalent. And this is the only requirement that I believe can be imposed on two correct analyses of the same concept.

It is indeed an interesting fact that the analysis of (1) in the system E leads to a result which is not intensionally isomorphic to the result of the analysis of (1′) in the system G even when the analyses are constructed on the same pattern. This is easily seen to be the result of the fact that (7) quotes a sentence of E, while (7′) quotes its translation in G, and the *names* of different intensionally isomorphic expressions are not intensionally isomorphic; in fact they are not synonymous in *any* sense. We could of course construct a sentence in the system E which would be intensionally isomorphic to (7′) (let us call it '(*c*)'), and a sentence in G which would be intensionally isomorphic to (7) (let us call it '(*d*)'); but I do not think we should consider it as a theoretical question: 'which is the correct analysis of (1)—(7) or (7′) or (*c*) or (*d*)?' If one is correct, then *all* are, and it does not matter that some are intensionally isomorphic and some are not.

In closing this part of my discussion, I should like to remark that if one does wish to emend Carnap's theory so that the analysis of (1) (or of (1) and (1′)) will lead to intensionally isomorphic results in the systems E and G one has only to specify that the quoted sentence should not be "Man is a rational animal" or "Der Mensch ist ein vernuenftiges Tier", but the translation of this sentence into an arbitrarily selected neutral system, say the system L, corresponding to Latin. Then in (7) the words ' "Man is a rational animal", as sentence of E' become replaced by the words ' "Homo est animal rationale", as sentence of L', and in (7′) the words ' "Der Mensch ist ein vernuenftiges Tier", als Satz von G' are replaced by ' "Homo est animal rationale", als Satz von L'; and (7) and (7′) are then intensionally isomorphic. But I do not think the advantages are sufficiently great to warrant this revision.

We are now in a position to consider Church's final criticism. Church points out that the result of prefixing 'John believes that' to (7) may have a different truth value from the result of prefixing 'John glaubt dass' to (7′). But this, like the remark that (7) and (7′) are "not intensionally isomorphic", is a crushing blow only if one has somehow been led to agree that (7) and (7′) *ought* to be synonymous.

II

Mates remarks:[7]

> Carnap has proposed the concept of intensional isomorphism as an approximate explicatum for synonymity. It seems to me that this is the best proposal that has been made by anyone to date. However, it has, along with its merits, some rather odd consequences. For instance, let "D"

and "D′" be abbreviations for two intensionally isomorphic sentences. Then the following sentences are also intensionally isomorphic:

(14) Whoever believes that D, believes that D.
(15) Whoever believes that D, believes that D′.

But nobody doubts that whoever believes that D believes that D. Therefore, nobody doubts that whoever believes that D believes that D′. This seems to suggest that, for any pair of intensionally isomorphic sentences—let them be abbreviated by "D" and "D′"—if anybody even doubts that whoever believes that D believes that D′, then Carnap's explication is incorrect.

This argument seems extremely powerful. Suppose, for the sake of illustration, that we use 'Hellene' as some newspapers do, as a synonym for 'Greek'. Then 'All Greeks are Greeks', and 'All Greeks are Hellenes', are intensionally isomorphic. Hence 'Whoever believes that all Greeks are Greeks believes that all Greeks are Greeks' and 'Whoever believes that all Greeks are Greeks believes that all Greeks are Hellenes' are intensionally isomorphic, and so (supposedly) synonymous. Now I do not myself doubt that 'Whoever believes that all Greeks are Greeks believes that all Greeks are Hellenes' is true; but it is easy to suppose that someone *does* doubt this, whereas it is quite likely that nobody doubts that whoever believes that all Greeks are Greeks believes that all Greeks are Greeks.
Accordingly:

(*e*) Nobody doubts that whoever believes that all Greeks are Greeks believes that all Greeks are Greeks.

and

(*f*) Nobody doubts that whoever believes that all Greeks are Greeks believes that all Greeks are Hellenes.

may quite conceivably have opposite truth value, and so *cannot* be synonymous.

Mates only suggests that this may invalidate Carnap's original proposal; Carnap, however, takes a harsher attitude toward his own theory; he believes that his theory in its present form cannot refute this criticism.

Mates goes on: "What is more, *any* adequate explication of synonymity will have this result, for the validity of the argument is not affected if we replace the words "intensionally isomorphic" by the word "synonymous" throughout.[8] In short, on any theory of synonymity, the synonymity of (*f*) and (*e*) must follow if the synonymity of 'Greek' and 'Hellene' is assumed. If we take this seriously, there is but one conclusion to which we can come: 'Greek' and 'Hellene' are not synonyms, and by the same argument, neither are any two different terms. This is a conclusion which some authors would be prepared to accept, even on other grounds."[9]

I believe, however, that the felt synonymity of such different expressions as 'snow is white' and 'Schnee ist weiss', or (in the use described above) of 'Greek' and 'Hellene', is undeniable. To maintain this synonymity involves our denying that in fact the synonymity of (*f*) and (*e*) follows from the synonymity of 'Greek' and 'Hellene'. Can this be denied?

At first blush, it would seem that it cannot. If two expressions have the same meaning, they can be interchanged in any context. We state this formally:

(g) The sense of a sentence is a function of the sense of its parts.

In a moment we shall criticize the apparent "self-evidence" of this principle. But first let us pause to make one point quite clear: whoever accepts (*g*) must conclude that no two different expressions have the same sense, for Mates's argument is formally sound if the interchange of expressions with the same sense is permitted in every context; conversely, whoever believes that two different expressions *ever* have the same sense must, by the same token, reject the principle. This would seem to be entirely destructive of some present theories of meaning-analysis which appear to involve simultaneous acceptance of (*g*) and the synonymity of some distinct expressions.

Let us now return to (g). 'The sense of a sentence is a function of the sense of its parts'. Let us ask why this seems to be self-evident. There is of course the formal similarity to the 'equals may be substituted for equals' which we learn in high school mathematics; but let us put this aside as irrelevant. We may suppose that if we were to ask someone who accepts this principle why it is true, he might well reply with the theoretical question 'Of what else could it be a function?' And just this is the heart of the matter.

Consider, for the moment, a simpler example (a variant of the famous "paradox of analysis"): 'Greek' and 'Hellene' are synonymous. But 'All Greeks are Greeks' and 'All Greeks are

Hellenes' do not *feel* quite like synonyms. But what has changed? Did we not obtain the second sentence from the first by "putting equals for equals"? The answer is that the *logical structure* has changed. The first sentence has the form 'All F are F', while the second has the form 'All F are G'—and these are wholly distinct (the first, in fact, is L-true, while the second schema is not even L-determinate). This suggests the following revision of the principle:

(*h*) The sense of a sentence is a function of the sense of its parts *and of its logical structure.*[10]

I believe that a large part of the "self-evidence" of (*g*) arises from the fact that we do not consider any alternatives: when, in particular, we contrast (*g*) and (*h*), I think that we find the latter principle even more plausible than the former. It is easy to illustrate the pervasive importance of logical structure as a factor in meaning: if it is through the names occurring in it that a sentence speaks about the world, it is through its logical structure that a sentence has implication relations to other sentences, and it is upon logical structure, or syntax, that the correctness of all our logical transformations depends.

The foregoing considerations lead us, therefore, to the following modification in the definition of intensional isomorphism:

(*i*) Two expressions are intensionally isomorphic if they have the same logical structure, and if corresponding parts are L-equivalent.

This amounts to saying that two expressions are intensionally isomorphic if (*a*) they are intensionally isomorphic in Carnap's sense, and (*b*) they have the same logical structure. It is proposed that "intensional isomorphism" so defined should be the explicans for synonymity in the strongest sense (interchangeability in belief contexts and indirect discourse).[11]

III

We must now consider another possible solution to the problem posed by Mates: that of Frege. According to Frege, a sentence in an "oblique" context (i.e. a belief context, or indirect discourse) does not have its ordinary nominatum and sense; rather it names the proposition that it normally expresses, i.e. that is normally its sense, and it expresses a new sense called its "indirect" or "oblique" sense:[12]

> . . . there is also . . . indirect discourse, of which we have seen that in it the words have their indirect (oblique) nominata which coincide with what are ordinarily their senses. In this case then the clause has as its nominatum a proposition, not a truth-value; its sense is not a proposition but it is the sense of the words 'the proposition that . . .'.

This view leads to the following answer to Mates's problem: even if D and D′ ordinarily have the same sense, in (14) and (15) they have their indirect senses, and they have their normal sense as nominatum. Hence (14) and (15) do not have the same sense but only the same nominatum (truth-value); and 'nobody doubts that (14)'[13] need not even have the same truth-value as 'nobody doubts that (15)', because in 'nobody doubts that (14)' the whole of (14) occurs in its indirect sense.

Mates's paradox is reinstated in a milder but still extremely damaging form for this theory, however, by considering the case in which D and D′ are two expressions with the same oblique sense. In such a case (*e*) and (*f*) would necessarily have to have the same truth-value; and we conclude by the same argument as before that no distinct expressions can ever have the same indirect sense.[14]

This appears to be a serious defect in Frege's theory. Frege himself certainly holds that different expressions may have the same sense in belief contexts. Thus he asserts in the passage quoted that the indirect sense of '. . .' is the same as the sense of the words 'the proposition that . . .', for example:

(*j*) John believes the earth is round

and

(*k*) John believes the proposition that the earth is round,

have the same sense. But some philosophers doubt that there are propositions, and hence that (strictly speaking) anyone ever *believes a proposition*. Such a philosopher would doubt that:

(*l*) If anyone believes the earth is round, he believes the proposition that the earth is round.

But he certainly would not doubt that:

(*m*) If anyone believes the earth is round, he believes the earth is round.

Therefore 'someone doubts (*l*)' does not have the same nominatum (truth-value) as 'someone doubts (*m*)', and accordingly the "indirect" sense of the 'the earth is round' is *not* the sense of the words 'the proposition that the earth is round'—contrary to Frege's assertion. But then it becomes difficult to say just what is the indirect sense of 'the earth is round'.

Thus we see that we cannot make the slightest change in the wording of a belief sentence without altering its sense. And it can now be shown that we cannot interchange different expressions in a *reiterated* oblique context (e.g. 'George believes that John says . . .') and hope to maintain even logical equivalence. For consider:

(*n*) Betty said that John is a Hellene

and

(*o*) Betty said that John is a Greek.

Since these have different senses, the result of prefixing 'John believes that' to (*n*) may even have a different truth-value from the result of prefixing 'John believes that' to (*o*).[15]

Let us imagine a case in which John says: 'Betty said that I am a Greek.' I believe that we should regard:

(*p*) John believes that Betty said that he is a Greek

and

(*q*) John believes that Betty said that he is a Hellene

as both constituting correct descriptions of this situation. Otherwise, it would appear, we are construing (*p*) as meaning "John believes that Betty said 'John is Greek' "; and this would amount to taking the quotation as a *direct* quotation, not an indirect one.

In any event, the case for regarding (*p*) and (*q*) as equivalent seems exactly as good as the case for regarding (*n*) and (*o*) as equivalent. To give up the equivalence of (*n*) and (*o*) would of course be to give up indirect quotation altogether; but to maintain it, while denying the equivalence of (*p*) and (*q*), is arbitrary.

In concluding, I should like to point out some applications of the concept of synonymy presented in this paper (in (*i*) above) to some classical problems of meaning analysis. In the first place, let us consider:

(*r*) George asked whether the property Greek is identical with the property Hellene and

(*s*) George asked whether the property Greek is identical with the property Greek.

This is clearly similar to Russell's "author of *Waverley*". But certain differences are relevant. The theory of descriptions will not take care of the problem posed by (*r*) and (*s*), for 'the property Greek' and 'the property Hellene' may well be designated by constants and not descriptions, even in *Principia Mathematica*. But in the theory presented above, no difficulty arises: 'The property Greek is identical with the property Greek' and 'The property Greek is identical with the property Hellene' are simply not synonymous (this is another instance of the "paradox of analysis"); hence (*r*) and (*s*) are not synonymous. The case presented by Carnap in his reply to Linsky,[16] of the sentences '5 is identical with 5' and '5 is identical with V', is disposed of in the same fashion.

Finally, let us look at Mates's sentences (14) and (15) above. If D and D' are different expressions, (14) and (15) are *never* synonymous, on our analysis. Thus we are not disturbed by the fact that 'Nobody doubts that (14)' may have a different truth value from 'Nobody doubts that (15)'.

Notes

1. A. Church '*Carnap's Analysis of Statements of Assertion and Belief,*' *Analysis* 10 (1950), 98–9.
2. R. Carnap, *Meaning and Necessity* (Univ. of Chicago Press, 1947).
3. '*Synonymity,*' *University of California Publications in Philosophy 25 (1950), 201–26, reprinted in Leonard Linsky (ed.). Semantics and the Philosophy of Language* (University of Illinois Press, 1952).
4. This refers to the explicans proposed for the concept of synonymity in *Meaning and Necessity*. It was Prof. Carnap who pointed out to me the significance of Mates's criticism, and this paper owes its existence to his stimulus and help.
5. The letters 'E' and 'G' will denote the semantical systems corresponding to English and German.
6. Cf. 'Carnap's Analysis', p. 99.
7. 'Synonymity', p. 215.
8. 'Synonymity', p. 215.
9. Cf. N. Goodman, 'On Likeness of Meaning', *Analysis*, 10 (1949).
10. Two sentences are said to have the same *logical structure*, when occurrences of the same sign in one correspond to occurrences of the same sign in the other.
11. For a distinction of stronger and weaker concepts of synonymity, see Carnap, 'Reply to Leonard Linsky'. *Philosophy of Science* 16. 4 (1949), 347–50.
12. Cf. G. Frege, 'On Sense and Nominatum', reprinted in Feigl and Sellars (eds) *Readings in Philosophical Analysis* (Appleton-Century-Crofts, 1949).

13. '(14)', '(15)', etc. are used as abbreviations here and in similar positions later.
14. This defect in Frege's theory was pointed out to me by Carnap.
15. The difference in sense could again be directly established by using our pattern 'nobody doubts that if (*n*) then (*o*)'.
16. Cf. 'Reply to Leonard Linsky'.

IX

INTENSIONAL ISOMORPHISM AND IDENTITY OF BELIEF

ALONZO CHURCH

I

THE criterion that beliefs expressed by given sentences are identical if and only if the sentences are intensionally isomorphic is contained in Carnap's analysis of belief statements,[1] and it may be advantageous to separate this criterion from other features of Carnap's analysis, in order to examine it independently.

For our present purpose it will be sufficient to confine attention to a single language, which we may take to be Carnap's S_1 with various individual and predicator constants added to it as required,[2] and to consider L-equivalence and intensional isomorphism, only of designator matrices[3] within this one language and containing the same free variables. It will be recalled that Carnap's definition of 'intensionally isomorphic' depends on a definition of the semantical term 'L-true'.[4] The designator matrices A and B, containing the same free variables, are then said to be L-equivalent if and only if the closure of $A \equiv B$ is L-true.[5] And two designator matrices containing the same free variables are said to be intensionally isomorphic if one can be obtained from the other by a series of steps which consist of (1) alphabetic changes of bound variable, (2) replacements of one individual constant by another which is L-equivalent to it, and (3) replacements of one predicator constant by another which is L-equivalent to it.[6]

To intensional isomorphism, in this sense, as criterion of identity of belief, there are objections which may be offered on the basis of Carnap's Principle of Tolerance, the principle namely that

Alonzo Church, 'Intensional Isomorphism and Identity of Belief', *Philosophical Studies* 5 (1954) 65–73.

everyone is at liberty to build his own form of language as he will[7].

By the Principle of Tolerance, no one shall forbid us to introduce two completely synonymous predicator constants, or two completely synonymous individual constants, into a language (such as Carnap's S_1), if we choose to do so. Exactly this situation is evidently contemplated in Carnap's definition of 'intensionally isomorphic', and published informal discussions of the definition have in fact sought out examples of synonymous constants, for example, in the English language, to be used for purposes of illustration. It is true that formalized languages constructed by logicians rarely contain synonymous primitive constants, as it is clear that the inclusion of such synonyms among the primitive constants would not be consistent with the logician's usual demand for economy of primitives. But to object to a language on the ground of lack of economy is not to say that it is an inadmissible language, but only that it fails to serve a certain purpose. (And the same language which fails to serve one purpose may for that very reason better serve another.)

However, by the Principle of Tolerance, it is also possible to introduce into a language like S_1 two predicator constants (or two individual constants) which are L-equivalent but not synonymous. For example, let the individuals be the positive integers, and let P and Q be predicator constants, such that Pn expresses that n is less than 3, and Qn expresses that there exist x, y, and z such that $x^n + y^n = z^n$. It is of course permissible to introduce P and Q as primitive constants, together perhaps with axioms containing them, such as may be suggested by their meanings.[8] For the sake of illustration let us suppose that Fermat's claim, to have had a proof of his (now so-called) Last Theorem, was correct. Then P and Q are L-equivalent, and it may even be possible to prove $(n)\,[\mathrm{P}n \equiv \mathrm{Q}n]$ from the axioms. Yet it is evident that one might believe that $(\exists n)\,[\mathrm{Q}n \sim \mathrm{P}n]$ without believing that $(\exists n)\,[\mathrm{P}n \sim \mathrm{P}n]$, since the proof of Fermat's Last Theorem, though it be possible, is certainly difficult to find (as the history of the matter shows).

Thus if intensional isomorphism is to serve as criterion of identity of belief, Carnap's definition requires the following admendment:

> In (2) and (3) as given above, the condition of L-equivalence shall be replaced by that of synonymy.

Again by the Principle of Tolerance it is possible to introduce a predicator constant which shall be synonymous with a specified abstraction expression of the form $(\lambda x)[..x..]$; or to introduce an individual constant synonymous with a specified individual description of the form $(\iota x)[..x..]$. And (unlike the case of synonymous primitive constants) it may be held that something like this actually occurs in formalized languages commonly constructed—namely those in which definitions are treated as introducing new notations into the object language,[9] rather than as metatheoretic abbreviations. But whether or not the process is called definition, it is clear by the Principle of Tolerance that nothing prevents us from introducing (say) a predicator constant R as synonymous with the abstraction expression $(\lambda x)[..x..]$, and taking $R \equiv (\lambda x)[..x..]$ as an axiom.[10] And if this is done, then R must be interchangeable with $(\lambda x)[..x..]$ in all contexts, including belief contexts, being synonymous with $(\lambda x)[..x..]$ by the very construction of the language—by definition, if we choose to call it that.

Thus we are led to a second amendment of Carnap's definition, as follows:

> In addition to (1), (2), and (3), as given above, steps of the following kinds shall also be allowed: (4) replacement of an abstraction expression by a synonymous predicator constant; (5) replacement of a predicator constant by a synonymous abstraction expression; (6) replacement of an individual description by a synonymous individual constant; (7) replacement of an individual constant by a synonymous individual description.

For intensional isomorphism as modified by these two amendments of Carnap's definition, let us introduce the name 'synonymous isomorphism'. It is proposed that synonymous isomorphism, as thus defined for the language S_1 and as extended by more or less obvious analogy to many other languages,[11] should replace Carnap's intensional isomorphism as criterion of identity of belief.

In order to make this possible, it is necessary to provide a determination of synonymy as part of the semantical basis of S_1 or of other language employed. This might be done directly, by means of *rules of synonymy* and rules of non-synonymy, or it might be done indirectly by means of *rules of sense*.[12] In either case there are certain obvious limitations upon the Principle of Tolerance which must be taken into account: for example, though we are at liberty in introducing a new constant to fix its meaning in any non-circular fashion that we please, and in particular to make it synonymous with any expression already at hand, we may not by arbitrary convention make the constant synonymous with an expression containing that same constant; and having once fixed the meaning of a constant, we are not then free to make further arbitrary conventions about its meaning (in particular, the same constant may not be made synonymous with two different expressions unless one of these synonymies can be shown to be a consequence of the other).[13]

II

Since our proposal of synonymous isomorphism is almost opposite in tendency to a modification of intensional isomorphism which is proposed in a recent paper by Hilary Putnam,[14] and which seems to be at least partly supported by Carnap,[15] it becomes necessary to consider Putnam's proposal, and in fact to rebut it (in the sense of showing it to be superfluous) if our own is to be maintained. Both Putnam and Carnap rely heavily on a brief remark in a paper of Benson Mates,[16] in such a way that it will be sufficient for our purpose if Mates's remark (as interpreted by Putnam) can be refuted

Mates introduces two sentences D and D′ which shall be particular sentences that are different but intensionally isomorphic. The two sentences D and D′ being not otherwise specified by Mates, let us choose them for the purpose of the present discussion as follows:

D. The seventh consulate of Marius lasted less than a fortnight.

D′. The seventh consulate of Marius lasted less than a period of fourteen days.

For the sake of the illustration, we suppose that the word 'fortnight', in English, means a period of fourteen days and is synonymous with 'a period of fourteen days'.[17] And in order to secure the complete synonomy of D and D′, we have used in D′ the phrase 'less than a period of fourteen days' rather than the shorter and more natural 'less than fourteen days'.

The sentences D and D′, as chosen above, are then not intensionally isomorphic, but synonymously isomorphic. They serve our present purpose the better for that very reason. In fact Mates, though directing his remark in the first instance against intensional isomorphism, concludes by saying that it is not affected if 'intensionally isomorphic' is replaced by 'synonymous' throughout.

And in reproducing Mates's argument we shall replace his 'intensionally isomorphic' everywhere by 'synonymously isomorphic'—synonymous isomorphism being our proposed explicatum of synonomy.

Consider then, following Mates, the two sentences:

(14) Whoever believes that the seventh consulate of Marius lasted less than a fortnight believes that the seventh consulate of Marius lasted less than a fortnight.

(15) Whoever believes that the seventh consulate of Marius lasted less than a fortnight believes that the seventh consulate of Marius lasted less than a period of fourteen days.

According to Mates, it is true that:

(16) Nobody doubts that whoever believes that the seventh consulate of Marius lasted less than a fortnight believes that the seventh consulate of Marius lasted less than a fortnight.

But it is not true that:

(17) Nobody doubts that whoever believes that the seventh consulate of Marius lasted less than a fortnight believes that the seventh consulate of Marius lasted less than a period of fourteen days.

In fact a counter-example against (17) is evidently provided by philosophers who have considered the question of the criterion of identity of belief, and perhaps in particular by readers of this paper. For by considering this question of philosophical analysis, one is almost inevitably led, at least tentatively, to doubt that (15), or else to entertain an analogous doubt in the case of some other pair of synonymously isomorphic sentences (in place of D and D′). Even if this doubt is afterwards overcome by some counter-argument, the very possibility of entertaining the doubt that (15), without simultaneously doubting that (14), shows (14) and (15) to be non-interchangeable in belief contexts.[18] The historical facts as to who has doubted what or as to the truth of (16) are not really relevant here,[19] but only the *possibility* of doubting that (15) without doubting that (14). Since, according to Mates, (14) and (15) are synonymously isomorphic, the result is to discredit synonymous isomorphism as the criterion of identity of belief.

It must be understood that those who are supposed to have doubted that (15) without doubting that (14) are supposed also to have had a sufficient knowledge of the English language so that the doubt was not, for example, a doubt about the meaning of the word 'fortnight' in English.

Nevertheless, it is natural to suggest as a means of overcoming Mates's difficulty that it is after all not possible to doubt that (15) without doubting that (14); and that the doubt which as been or may have been sometimes entertained by philosophers in considering the question of the criterion of identity of belief is not the doubt that (15), but a doubt that does have reference to linguistic matters, namely the doubt that:

(18) Whoever satisfies in English the sentential matrix '*x* believes that the seventh consulate of Marius lasted less than a fortnight' satisfies in English the sentential matrix '*x* believes that the seventh consulate of Marius lasted less than a period of fourteen days'.[20]

If this suggestion can be supported, the difficulty urged by Mates disappears, as (18) is clearly not synonymously isomorphic either to (14) or to:

(19) Whoever satisfies in English the sentential matrix '*x* believes that the seventh consulate of Marius lasted less than a fortnight' satisfies in English the sentential matrix '*x* believes that the seventh consulate of Marius lasted less than a fortnight'.[21]

Now the test of translation into another language, originally suggested by C. H. Langford, is often valuable in determining whether a statement under analysis is to be regarded as a statement about some sentence, linguistic expression, or word, or rather as about something which the sentence, expression, or word is being used to mean.[22] I have used this test elsewhere to support the conclusion that the object of a belief shall be taken to be a proposition rather than a sentence, if certain important features of the ordinary usage of indirect discourse are to be preserved.[23] But I say that the same test in the present connection leads to a conclusion of opposite kind—namely that the doubt whose existence or possibility Mates urges (as a difficulty in the analysis of belief statements) is a doubt about certain sentential matrices, and thus a doubt that (18) rather than a doubt that (15).[24]

Let us therefore translate (14), (15), and (18) into German. The translation of (18) is:

(18) Wer auf Englisch die Satzmatrix '*x* believes that the seventh consulate of Marius lasted less than a fortnight' erfüllt, erfüllt auf Englisch die Satzmatrix '*x* believes that the seventh consulate of Marius lasted less than a period of fourteen days'.

As soon as we set out to translate (14) and (15), our attention is drawn to the fact that the German language has no single word which translates the word 'fortnight', and that the literal translation of the word 'fortnight', from English into German is 'Zeitraum von vierzehn Tagen'.[25] In consequence, the German translations of (14) and (15) are identical, as follows:

(14′) (15′) Wer glaubt dass das siebente Konsulate des Marius weniger als einen Zeitraum von vierzehn Tagen gedauert habe, glaubt dass das siebente Konsulat des Marius weniger als einen Zeitraum von vierzehn Tagen gedauert habe.

Of course we must ask whether the absence of a one-word translation of 'fortnight' is a deficiency of the German language in the sense that there are therefore some things which can be expressed in English but cannot be expressed in German. But it would seem that it can hardly be so regarded—else we should be obliged to call it a deficiency of German also that there is no word to mean a period of fifty-four days and six hours, or that the Latin word 'ero' can be translated only by the three-word phrase 'ich werde sein'. Indeed it should rather be said that the word 'fortnight' in English is not a necessity but a dispensable linguistic luxury.

Granted this, let us translate into German 'Mates doubts that (15) but does not doubt that (14).[26] As the resulting German sentence is a direct self-contradiction, and as it cannot matter to the soundness of our reasoning whether we carry it out in English or in German, we must conclude that Mates (whatever he himself may tell us) does not really so doubt—and that he must have mistaken the doubt that (18) for the doubt that (15).

Notes

1. R. Carnap, *Meaning and Necessity* (University of Chicago Press, 1947), §§13–15.
2. We also suppose that Carnap's sign ' ≡ ' of identity of individuals is a predicator constant, and that when A and B are individual expressions, A ≡ B is to be understood merely as an abbreviation or alternate way of writing ≡ AB. This modification of S_1 serves to simplify the discussion but is not otherwise essential to the conclusions we reach.
3. I follow Carnap's terminology, in spite of my own preference for a somewhat different terminology—e.g. 'well-formed formula' instead of 'designator matrix'.
4. The definition of 'L-true' need not be repeated here. But notice should be taken of two necessary corrections to the definition as it is developed in §§1–2 of Carnap's book.

   In 2-2 the correction of Kemeny must be adopted (*Journal of Symbolic Logic*, 16 (1951), 206); i.e. in place of "every state-description" the restriction must be made to non-contradictory state-descriptions. Otherwise consequences will follow that are certainly not intended by Carnap, for instance that no two different atomic sentential matrices (and no two different predicator constants) can be L-equivalent.

   In the rules of designation 1-1 and 1-2, the way in which the English language and certain phrases of the English language are mentioned, rather than used, is inadmissible—as may be seen by the fact that it forces the tacit use, in 1-3 and 1-4, of certain rules of designation *of the English language*, which, if stated, would have a quite different form from 1-1 and 1-2. For example, Carnap's rule of designation, " '*s*' is a symbolic translation [i.e. from English] of 'Walter Scott'", should be changed to a rule which mentions the man Walter Scott rather than the words 'Walter Scott'; perhaps it should be simply "'*s*' refers to Walter Scott", in order to justify the inference from 1-3 to 1-4.

   These corrections are not directly relevant to the present paper, but our discussion presupposes that suitable corrections have been made.
5. Carnap uses ' ≡ ' not only between sentential matrices as a sign of material equivalence, but also between other designator matrices as a sign of identity (in place of the usual ' = ').
6. Because of the restriction to the single language S_1 and to designator matrices containing the same free variables, we have been able to give a simplified form to Carnap's definitions of 'L-equivalent' and 'intensionally isomorphic'.
7. In this form, as applied to the construction of a new language and the determination of what its expressions shall mean, the Principle of Tolerance is hardly open to doubt. The attempt to apply the Principle of Tolerance to the transformation rules of a language after the meaning of the expressions of the language has already been determined (whether by explicit semantical rules or in some looser way) is another matter, and certainly doubtful, but is not at issue here. In fact Carnap (if he ever did) does not now maintain the Principle of Tolerance in this latter and more doubtful form (see §39 of his *Introduction to Semantics* (Harvard University Press, 1942)).
8. There is no condition to the effect that a predicator constant must express a simple property, rather than such a comparatively complex property as that which is here expressed by Q. In fact some of Carnap's examples of predicator

constants express properties which are evidently not especially simple. And it is, moreover, not clear how the distinction between a simple and a complex property could be made precise in any satisfactory way (except by making it relative to the choice of a particular language).

9. This is the account of definition which is given, for example, by Hilbert and Bernays in *Grundlagen der Mathematik*. In constructing formalized languages, others (including myself) have often preferred to avoid definitions in this sense, which change the object language by adding new notations to it. But such avoidance is on the same ground of economy that underlies the avoidance of synonymous primitive constants, and need not be demanded when economy is not the objective.
10. Compare Carnap, *The Logical Syntax of Language*, (Routledge and Kegan Paul, 1937), §22, 1(*b*).
11. In particular to any of the languages considered in my paper, 'A Formulation of the Logic of Sense and Denotation', (*Structure, Method and Meaning*, pp. 3–24), and to languages obtained from these by adding constants of any types, with specified meanings.

    It is necessary to explain that the statement on page 5 of that paper, that Alternative (0) "may be described roughly by saying that it makes the notion of sense correspond to Carnap's notion of intensional structure" is an error (unless "roughly" is understood in a very liberal sense). The intention of Alternative (0) is rather that two well-formed formulas shall have the same sense if and only if they are synonymously isomorphic.
12. See my 'The Need for Abstract Entities in Semantic Analysis'. *Proceedings of the American Academy of Arts and Sciences* 80, 1 (1951), 100–12.
13. Compare the "rules of definition," originally Aristotelian, which are often included in books on traditional logic.
14. 'Synonymy, and the Analysis of Belief Sentences', *Analysis* 14. 5 (954), 114–22, also in this volume.
15. In a forthcoming paper 'On Belief-Sentences: Reply to Alonzo Church'.
16. Synonymity', *University of California Publications in Philosophy* 25 (1950), see the lower half of page 215.
17. To treat the English language as a language for which syntactical and semantical rules have been fully given is of course to make a supposition contrary to fact, but it is one which is very convenient for illustrative purposes and has in fact been adopted in informal discussion by Carnap, Mates, Putnam, and many others. Use of this device has the effect that it may be necessary in the course of the illustration just to invent a rule of English, either to fill a gap in the rules as found in existing grammars and dictionaries or to remove an equivocacy. In the present context, for instance, we have been obliged to decide arbitrarily (or on the basis of mere plausibility) that 'fortnight' is synonymous with 'a period of fourteen days' rather than with 'a period of two weeks'; existing English dictionaries either fail to decide this point or disagree among themselves, probably because universal familiarity with the multiplication table tends to obscure the fact that the two later (quoted) phrases are not synonymous with each other.
18. A context of doubting is of course a belief context, since to doubt is to withhold belief. And a criterion of identity of belief must also be a criterion of identity of doubt.
19. Doubt being one of the fundamentals of philosophical method, it would be

difficult indeed to find a proposition that some philosopher might not be found to doubt.

20. The two occurrences of the phrase 'in English' would usually be omitted, but strictly they are necessary; for the semantical relation of satisfaction (or fulfilment) is a ternary relation among an individual, a sentential matrix, and a language.
21. The point is that *names* of two different sentences are not synonymous in any sense, and in particular not synonymously isomorphic, even though the sentences themselves be synonymously isomorphic.
22. (Added August 4, 1954). The existence of more than one language is not usually to be thought of as a fundamental ground of the conclusions reached by this method. Its role is rather as a useful device to separate those features of a statement which are essential to its meaning from those which are merely accidental to its expression in a particular language, the former but not the latter being invariant under translation. And distinctions (e.g. of use and mention) which are established by this method it should be possible also to see more directly. The point is well illustrated by a paper of Wilfrid Sellars, 'Putnam on Synonymity and Belief', forthcoming in *Analysis*, in which conclusions the same as or similar to those of Part II of this paper are reached by a more direct analysis. Professor Sellars's paper and mine were written independently, but I saw a copy of it by return mail when my own was submitted to *Philosophical Studies*.
23. *Analysis*, 10. (1950), 97–9.
24. The object of the doubt must still be a proposition, but a proposition about certain sentential matrices.
25. The shorter translation 'vierzehn Tage' would be more usual, but is not quite literal, as may be seen by considering the question of translating the phrase 'three fortnights' into German.
26. Of course '(14)' and '(15)' are here used, not as names of the sentences which we have so numbered, but just as convenient abbreviations. The reader must imagine the full sentences written out in place of the '(14)' and '(15)'. Indeed, throughout the paper such parenthetic numerals are to be understood as *abbreviations* when preceded by the word 'that'—but elsewhere as *names* of their sentences.

X

DIRECT REFERENCE AND ASCRIPTIONS OF BELIEF

MARK RICHARD

It is often supposed that demonstratives and indexicals are *devices of direct reference*—that they are, as David Kaplan puts it, terms which "refer directly without the mediation of a Fregean *Sinn* [or individual concept, set of properties, etc.] as meaning".[1] Most of the resistance to this view, I think, arises from the suspicion that it is not possible to give an acceptable treatment of the semantics of belief ascriptions and other so-called propositional attitude contexts which is consistent with the thesis of direct reference. For it seems that a straightforward construal of the thesis (along with some plausible semantical assumptions) requires that demonstratives, when co-referential, be intersubstitutable everywhere, even in belief contexts, *salva veritate*. But many feel that it is *obvious* that such substitutions do not always preserve truth.

The purpose of this paper is to motivate and present a semantics for a first-order treatment of belief ascriptions which is both consistent with the thesis of direct reference and intuitively satisfactory. The paper is structured as follows: In Section I, I discuss semantical consequences of the thesis of direct reference—in particular, what it does and does not require with respect to the overall form of a semantical treatment of belief ascriptions. I also discuss a view about belief, championed by Kaplan and John Perry, which I call the triadic view of belief. Crudely put, it is the view that belief is a triadic relation among a person, a proposition, and a sentential meaning, the latter entity a different sort of thing than a proposition. On this view, to believe a proposition is to do so 'under' a sentential meaning.

Mark Richard, 'Direct Reference and Ascriptions of Belief', *Journal of Philosophical Logic* 12 (1983), 425–52.

The champions of the triadic view of belief have shied away from using the view to motivate a semantic account of belief ascriptions.[2] But the triadic view of belief suggests, as I note at the end of Section I, that ascriptions of belief not only imply that a proposition is an object of belief, but that it is believed in a certain way. The purpose of Sections II and III of the paper is to show that an account of belief ascriptions, on which they behave in just this way, can be formalized rather easily, and that it nicely handles certain cases which, at first blush, seem quite problematic for those who accept the view that demonstratives are directly referential. Section II is concerned with the semantics of ascriptions of belief *de se*: that is, with giving a first-order syntax and semantics which adequately represents (different readings of) ascriptions of the form ⌜*a* believes himself to be *F*⌝ (and of allied forms) and the relations of such ascriptions to *de re* ascriptions of belief. Section III is concerned with semantics for 'standard' belief ascriptions in which the sentential complement to 'believes that' contains demonstrative and indexical terms.

I

The core of the thesis that demonstratives are directly referential is negative.[3] A somewhat long-winded way of expressing the thesis is this: associated with well-formed expressions of English (taken relative to a context) is an entity which we may call the expression's *content*. (As will become clear, I am using this expression, as well as the term 'proposition', in a very technical and circumscribed way.) Contents play a number of semantic roles. One of the things the content of a singular term does is to determine, relative to a possible circumstance of evaluation, an individual; the individual so determined at a circumstance, by the content of a term (taken relative to a context), is the referent of the term (taken relative to the context) at the circumstance.[4] To say that a term is directly referential is to say something about how its content determines an individual: it is to say that it does *not* do this by means of a complex of properties, a Fregean sense, an individual concept, etc. A picturesque way of putting the matter is this: the content of a directly referential expression, taken relative to a context, *is* that thing which the expression, taken relative to the context, has as a referent at any circumstance of evaluation.

What specific semantical consequences we suppose the thesis of direct reference to have depends, of course, upon what sort of semantical assumptions we make, beyond the assumption that demonstratives are directly referential. As noted at the beginning of this paper, it is often assumed that one consequence of the thesis of direct reference is that any ascriptions of belief which differ only in that one contains a demonstrative *d* where the other contains a demonstrative *d'* have the same truth value relative to any context in which *d* and *d'* are co-referential.

Those who think that this is a consequence of the thesis of direct reference seem to reason as follows: the content of a declarative sentence, taken relative to a context—I shall use the term *proposition* for sentential contents—is what determines the truth value of a sentence, relative to the context. Content, however, is determined functionally. More precisely: the content of a sentence in a context—what proposition it expresses—is a function solely of the contents of the parts of the sentence (in the context) and the syntax of the sentence. Now the thesis of direct reference surely requires that demonstratives that denote the same thing have the same content. Hence, given this thesis, one must say that *any* sentences, differing only with respect to co-referential demonstratives, express the same proposition and, therefore, have the same truth value.

An advocate of the view that demonstratives are directly referential *could*, I suppose, deny that co-referential and directly referential terms have the same content, where the content of a term is characterized as it was above. But such a denial, I think, makes the claim that a certain sort of expression is directly referential, extremely mysterious. For recall that the claim that a term is directly referential is the claim that the term does not have, as a content, a sense, an individual concept, etc. This, coupled with the fact that such expressions are supposed to behave as rigid designators (i.e. a use of such a term has the same referent at every circumstance of evaluation at which it has any referent at all), makes it difficult to see how one could coherently maintain that co-designative directly referential terms can have different contents. How, one wants to know, could they differ?

A more plausible response to the above argument, I think, is to deny that it is invariably the case that the content of a complex expression is a function of the contents of its parts and their syntactic mode of combination. For there is no reason to think that content is the only sort of semantic value which expressions may have. Indeed, those who subscribe to the thesis of direct reference generally recognize at least one other sort of semantic value which expressions have. Thus, for example, Kaplan holds that the linguistic meaning of an expression (meaning in the sense of what is known by one who understands an expression) is to be identified, not with its content, but with what Kaplan calls the expression's *character*, the function which takes a context to the content of the expression therein. If the only sort of semantic value we recognized was content, it would be strange, to say the least, to suppose that the determination of content was not functional across the board. However, once we recognize yet another sort of semantic value, it is far from clear that we should suppose that the content of the whole is invariably a function of the contents of the parts.

This is particularly unobvious when the second semantic value is related to content as is character: in a fairly clear sense, character determines content, not the converse. Should the language contain an operator which is sensitive (not just to content, but) to character, it is at least an open question as to whether content is invariably functional. For suppose there were an operator O such that $O(A)$ is true only if the character of A has P. Since expressions with distinct characters may, relative to a context, have the same content, it is at least *a priori* possible that there be expressions B and B' such that the character of B has P, the character of B' does not, but, relative to some context c, B and B' have the same content. In this case, $O(B)$ and $O(B')$ have different contents (i.e. express different propositions) relative to c, since they diverge in truth value.[5]

I contend that this is not just an idle possibility: in Section III, I will urge that 'believes that' is sensitive to linguistic meaning (construed as character) as well as to content. This, however, is anticipating matters somewhat. For the moment, all we have established is this: the advocate of direct reference need not assent to the view that ⌜a believes that S⌝ and ⌜a believes that S'⌝, taken relative to a context in which S and S' express the same proposition, invariably have the same truth value.

It is worth observing that the above argument is consistent with the claim that, in a circumscribed but nonetheless significant number of cases, the content of an expression is determined as a function of the contents of its parts. In particular, we have given no reason to think that the advocate of direct reference would deny that what proposition is expressed by a sentence which contains at most truth functional or simple intensional operators (i.e. temporal or modal operators) is determined functionally by the contents of its parts and their mode of combination. I shall assume, for the purposes of this paper, that an advocate of the view that demonstratives are directly referential would say this.

We are now ready to discuss belief and ascriptions of belief. A fairly 'standard' view of belief is that propositions, characterized as above, play the role of objects of belief in two senses: (*a*) they are objects of belief in the sense that belief is a dyadic relation, the second term of the relation being a proposition; (*b*) they are objects of 'belief' in the sense that an ascription of belief ⌜*a* believes that *S*⌝ is true iff what *a* denotes bears the belief relation to the proposition expressed by *S*.

Many who are sympathetic to the thesis of direct reference—notably Kaplan and John Perry[6]—have proposed that propositions are objects of belief in some sense, but that the relation between a person and a proposition, when the latter is an object of his belief, is somewhat more complex than the above account suggests. On this view—the triadic view of belief, as I will call it—belief is a triadic relation between a person, a sentential meaning (understood as being a Kaplanesque character), and a proposition; to believe a proposition is to do so under a sentential meaning.[7] I will use the term *acceptance* for the relation which one bears to a sentential meaning when one believes a proposition under it; it is to be understood that, on the triadic view, a proposition *p* is an object of someone's belief if and only if he accepts a meaning which, relative to the context of which he is the agent, has *p* as value.

It is not my purpose here to defend this view of belief. I will, however, note that a view of belief along these general lines seems mandatory for those who accept the thesis that demonstratives are directly referential and take propositions to be, in some important sense, objects of belief. For, to take an example, on the thesis of direct reference (making, of course, the kind of semantical assumptions we are currently making), someone who expresses something he believes by saying 'You [person *X* is addressed, say, through the telephone] are happy, but she [*X*, who is standing across the street, is demonstrated] is not happy' expresses belief in the same proposition as does one who addresses *X* and says 'You are happy, but you are not happy'. Without invoking a view like the triadic view, it is difficult to explain, or even explain away, the intuition that an irrationality is present in the latter belief which is not present in the former—for the object of belief, in the sense of proposition believed, is the same in both cases. Invoking the view, however, what one can say is this: what is irrational is not to have the proposition in question as an object of belief, but to believe it in the way the second person does. For a rational person who understands English must know that the second meaning cannot yield a truth.[8]

It is worth noting that on such an account of belief, propositions are not (or, at least, need not be) simply vestigial remains of the simpler dyadic view of belief, playing no particularly important or indispensible role in the triadic theory. Propositions are here identified with the *contents* of belief; meanings are identified with *manners* in which one may hold beliefs. Presumably, *what* is believed is just as important as *how* it is believed. Furthermore, it is, presumably, in terms of propositions (or, at least, mostly in terms of propositions) that we evaluate claims concerning the retention of belief and claims that two people have the same belief.

As I have characterized it, the triadic theory is a (metaphysical) theory about the nature of belief; as such, it is not directly concerned with the semantical problem of truth conditions for ascriptions of belief. Some partisans of the triadic view[9] have suggested that even though belief is to be understood as above, an ascription ⌜*a* believes that *S*⌝, *a* a term, *S* a sentence, is true exactly if *a*'s referent believes (i.e. has as the content of a belief) the proposition expressed by *S*. None the less, if we suppose the triadic view of belief to be correct, it is natural to think that the triadic nature of belief might be reflected in ascriptions of belief. One way in which such a reflection might occur is this: some ascriptions of belief may imply, not only that a particular proposition is an object of belief, but that it is believed in a certain way—that is, that it is believed under a meaning of a certain sort.

It is the task of the remainder of this paper to make it plausible that this indeed *is* true of ascriptions of belief. Of course, the account of belief ascriptions to be developed will be consistent with the thesis of direct reference as we have characterized it. We shall begin by developing, in the next section, a treatment of so-called *de se* ascriptions of belief.

II

A *de se* ascription of belief is one of the form of

(1) *a* believes himself to be *F*

where *a* is a term, and *is F* is a predicative expression.[10] It is widely acknowledged that a *de se* ascription of the form of (1) is not implied by the corresponding *de re* ascription, that of the form of

(2) There is an *x* such that *x* is identical with *a* and *x* believes that *x* is *F*.

I presume that the reader is acquainted with the standard arguments that sentences of the form of (2) do not imply ones of the form of (1);[11] I will assume here that such an implication does indeed fail. It is, initially, difficult to see how an advocate of the thesis of direct reference could deny that this implication fails.

The problem is this: consider a particular *de se* ascription—say, 'John believes himself to be wise'—and the corresponding *de re* ascription, 'There is an *x* such that *x* is John and *x* believes that *x* is wise'. Under what conditions is the *de se* ascription true? Presumably, it is used to ascribe to John belief in the proposition John would express by saying 'I am wise'; thus, one thinks, it will be true iff John believes that proposition. But on the thesis of direct reference, this is the proposition that is expressed by a sentence of the form ⌜*d* is wise⌝, *d* a directly referential term denoting John. It seems that the advocate of direct reference ought to hold that John's believing such a proposition is both necessary and sufficient for the truth of the corresponding *de re* ascription. After all, '$(\exists x)$ $(x =$ John and x believes that x is wise)' seems to ascribe to John a belief in what '*x* is wise' expresses when John is assigned to '*x*'; but the free variable, under an assignment, seems to be the paradigm of a directly referential term.

A straightforward account of why the implication does not go through is motivated by the triadic theory of belief. For one can say that the ascription 'John believes himself to be wise' implies that John believes the proposition that he would express by saying 'I am wise' in a certain way—roughly, under the meaning of 'I am wise'. Pretty obviously, given the triadic theory of belief, John can believe this proposition under meanings other than that of 'I am wise': perhaps, for example, John sees a reflection of himself, does not know that he sees himself, accepts the meaning of 'He [John demonstrates his reflection] is wise', but does not accept the meaning of 'I am wise'. As long as John believes the proposition under some meaning, the *de re* ascription is true. Thus, we have an explanation of how it is that the *de re* ascription fails to imply the *de se* ascription.

A general treatment of *de se* ascriptions may be developed along the following lines. First, let us introduce some structure to meanings. Instead of thinking of a meaning as simply a function from contexts to propositions, think of it as a pair $\langle\langle s_1, \ldots, s_n\rangle, M^n\rangle$ $(n \geqslant 0)$, where each s_i is a (demonstrative) term-meaning—a function from contexts to individuals—and M^n is an n-place predicate-meaning—a function from contexts to n-place properties. (I will, for the sake of expediency, identify n-place properties with functions from n-tuples of possible individuals to sets of possible worlds; propositions with zero-place properties—i.e. sets of worlds.) The proposition such a meaning yields in a context c, is, of course, the proposition p such that w is in p exactly if w is in $[M^n(c)](\langle s_1(c), s_2(c), \ldots, s_n(c)\rangle)$.

Note, now, that we can 'partially interpret' such meanings, relative to a context. For example, if we start with a meaning $m = \langle\langle s_1, s_2\rangle, M^2\rangle$ and a context c, we can 'plug in' the values of s_1 and M^2 in c to get a 'reduced meaning' $m' = \langle\langle s_2\rangle, P^1\rangle$, P^1 the one-place property such that $w \in P^1(u)$ iff $w \in [M^2(c)](\langle s_1(c), u\rangle)$. The reduced meaning m', in turn, corresponds to the function from contexts to propositions which, applied to a context c', yields the proposition that the value of s_2 in c' has P^1.

The basic intuition behind the general treatment of *de se* ascriptions we propose is this: A *de se* ascription

(3) *a* believes himself to be *F*

is true exactly if *a*'s referent believes the proposition that he is *F* (i.e. the proposition that he has the property which is expressed by ⌜is *F*⌝ relative to the context at which we interpret (3)) under a meaning m, which has as one of its reduced meanings $\langle\langle\{I\}\rangle, F\rangle$, where $\{I\}$ is the meaning of '*I*'. This, in turn, will be true precisely if *a*'s referent accepts a meaning which is the meaning of a sentence of the form ⌜$\phi(I)$⌝, where $\phi(x)$ expresses, relative to his context, the property F. When someone believes a proposition under such a meaning, we will say that he self-attributes the property, allowing us to state our view in summary form as: (3) is true exactly if *a*'s referent self-attributes the property expressed by ⌜is *F*⌝.[12]

The semantical details of a formalization of our approach to *de se* ascriptions are not particularly complex. Syntactical details, however, are slightly subtle.

Consider, to begin with, the behaviour of 'believes' in *de dicto* and *de re* ascriptions and in *de se* ascriptions. In the first two sorts of ascriptions, the belief operator—use 'B^r' to represent it—appears to operate on an *n*-place predicate ($n \geqslant 0$) to yield an $n + 1$-place predicate. For example, 'at the level of logical form', 'B^r' combines with 'x loves y' to yield '$zB^r(x$ loves $y)$'. The belief operator in *de se* ascriptions, on the other hand—let us use 'B^s' to represent it—apparently combines with an *n*-place predicate ($n > 0$) and a specification of an argument place to yield an *n*-place predicate. Thus, for example, applying 'B^s' to 'x loves y' and specifying the first argument place seems to yield something along the lines of 'zB^s (he himself loves y)'.

Of course, given that we do not want *de se* ascriptions to be implied by the corresponding *de re* ascriptions, we cannot assume that something like 'zB^s (he himself loves y)' is reducible to an expression involving 'B^r' and other syntactic operations. For example, we would not want to identify 'zB^s (he himself loves y)' with the result of applying the operation 'identifying the first two argument places' to 'zB^r (x loves y)'. For the latter object— '$zB^r(z$ loves $y)$'—will be true, relative to an assignment f, precisely if $f(z)$ believes *de re*, with respect to $f(z)$ and $f(y)$, that the former loves the latter.

Thus, we will use two distinct belief operators, 'B^r' and 'B^s', in our formalization. 'B^r' will, as is usual, take a sentential complement. We will, however, have 'B^s' take as complement a 'property abstract' (something of the form ⌜$\hat{x}(\phi)$⌝, ϕ a sentence). The reasons for treating 'B^s' in this way have, for the most part, to do with elegance in presentation. We could, in principle, allow 'B^s' to take a sentential complement, so long as we introduced apparatus for indicating what argument positions in an embedded sentence are 'specified argument places' in the sense indicated above. Such a treatment, however, is messier than need be.

It should be stressed that the decision to treat the *de se* belief operator in this way does *not* constitute surrender of the view that the objects of belief (i.e. the contents of belief, in the sense of

Section I) are uniformly propositions, nor does it make it at all inappropriate to say that something of the form ⌜$\alpha B^s \hat{x}(\phi)$⌝ is (a representation of) an ascription of *belief*. Our semantics will take a formula of the form of ⌜$\alpha B^s \hat{x}(\phi)$⌝ to be true precisely if α's referent believes a *proposition* under a meaning m which has $\langle\overline{\langle\{I\}\rangle, \hat{x}(\phi)}\rangle$ as a reduced meaning, where $\overline{\hat{x}(\phi)}$ is the property the semantics associates with $\hat{x}(\phi)$. Furthermore, as we will show, a *de se* ascription will, in this treatment, imply its corresponding *de re* ascription (and thus imply that a certain proposition is believed), although the converse implication, of course, will not hold.

The vocabulary and formation rules for our treatment are as follows. As primitive vocabulary items we have: A denumerable set $V = \{x_1, x_2, \ldots\}$ of variables; denumerable sets $Y = \{y_1, y_2, \ldots\}$ and $T = \{t_1, t_2, \ldots\}$ of demonstrative terms (used to represent, respectively, uses of second person singular 'you' and third person singular demonstratives such as 'he', 'she', 'that', etc.); the singular term: I; for each n, a denumerable set F^n of n-place predicates; the truth functors: ¬, ∧, ∨, →, ↔; the belief predicates: B^r, B^s; the abstraction operator: ^; the quantifiers: ∃, ∀; and, as punctuation, '(', ')'. We use D to name the set of demonstratives of the language, the set $Y \cup T \cup \{I\}$; $\mathscr{T}$, the set of terms, is $D \cup V$.

The definition of well-formed formula is:

1. If $\Pi \in F^n$ and $\alpha_1, \ldots, \alpha_n \in \mathscr{T}$, then ⌜$\Pi^n\ \alpha_1 \ldots \alpha_n$⌝ is a formula.
2. If ϕ and Ψ are formulas, then ⌜$\neg(\phi)$⌝, ⌜$(\phi) \wedge (\Psi)$⌝, ⌜$(\phi) \wedge (\Psi)$⌝, ⌜$(\phi) \rightarrow (\Psi)$⌝, and ⌜$(\phi) \leftrightarrow (\Psi)$⌝ are formulas.
3. If ϕ is a formula, $\alpha \in V$, then ⌜$\exists\alpha(\phi)$⌝, and ⌜$\forall\alpha(\phi)$⌝ are formulas.
4. If ϕ is a formula, $\alpha \in \mathscr{T}$, then ⌜$\alpha B^r(\phi)$⌝ is a formula.
5. If $\alpha \in \mathscr{T}$ and Γ is a proper abstract, then ⌜$\alpha B^s \Gamma$⌝ is a formula, where a proper abstract is any expression of the form ⌜$\hat{\alpha}(\phi)$⌝, ϕ a formula, and α a member of V which occurs freely in ϕ.
6. These are all the formulas.

Before discussing the semantics, it is perhaps worthwhile to discuss the intuitive readings of those expressions which are formulas in virtue of clauses (4) and (5). Consider the following well-formed expressions of the language:

(4) $IB^r(IB^r(FI))$

(5) $IB^r(IB^s\hat{x}(Fx))$

(6) $IB^s\hat{x}(xB^r(FI))$

(7) $IB^s\hat{x}(IB^r(Fx))$

(8) $IB^s\hat{x}(xB^s\hat{x}(Fx))$.

The semantic differences among these can be brought out as follows. Read 'Fx' as 'x is wise'; imagine me to be standing by a mirror. (4) through (8) can be understood as representing different readings of "I believe that I believe that I am wise", the difference in readings corresponding to different sorts of meanings under which I might hold the belief: (4) will be true simply if I hold a belief under the meaning of 'He believes [*de re*] that he is wise' (occurrences of 'he' always accompanied by a demonstration by me of my reflection); (5) corresponds to belief under the meaning of 'He believes himself to be wise'; (6) under 'I believe that he is wise'; (7), 'He believes that I am wise'; (8), 'I believe myself to be wise'.[13]

We define an interpretation for the language as a quartet $M = \langle U, W, C, V\rangle$ which obeys the following strictures:[14]

1. U, W, and C are non-empty and disjoint sets (which, intuitively, represent possible individuals, worlds, and contexts, respectively).
2. (*a*) Associated with each member c of C is four-tuple $\langle c_A, c_W, c_Y, c_T\rangle$,
   (i) $c_A \in U$ (c's agent)
   (ii) $c_W \in W$ (c's world)
   (iii) c_Y and c_T are denumerable sequences of members of U (the potential addressees and demonstrata of c).
   (*b*) $c = c'$ iff $c_A = c'_A$, $c_W = c'_W$, $c_Y = c'_Y$, and $c_T = c'_T$.
   (*c*) No world contains distinct contexts with the same agent.
3. V is a function which assigns
   (*a*) a member of $((\mathcal{P}(W))^{U^n})^C$ to each member of F^n, for each n;
   (*b*) sets of meanings to each member of C, where a meaning is a pair $\langle\langle s_1, \ldots, s_n\rangle, M^n\rangle$ $(n \geq 0)$, each $s_i \in U^C$ and M^n a member of $((\mathcal{P}(W))^{U^n})^C$.

A word on the workings of V is perhaps in order here. V's assignments to predicate letters are, intuitively, predicate-meanings (taken to be functions from contexts to properties). V's assignments to contexts are to be understood as representing the class of meanings under which the agent of the context holds beliefs; in the terminology of Section I, $V(c)$ is the set of meanings which c_A accepts. Note that, for each context c, $V(c)$ determines a set of propositions, a proposition p being in the set so determined by $V(c)$ exactly if, for some m in $V(c)$, m, completely interpreted relative to c, yields p. These of course, are the propositions which are objects of belief of the agent of c.

We must, in order to give a definition of truth, characterize the conditions under which the agent of a context self-attributes a property. This we do using the notion of a reduced meaning, introduced above. Where $M = \langle\langle s_1, \ldots, s_n\rangle, M^n\rangle$ is a meaning, a reduced meaning corresponding to M, relative to a context c, is any function in $\mathscr{P}(W)^c$ which results (in the way indicated above) by interpreting M^n and one or more of the s_i, relative to c. An i-reduced meaning is any reduced meaning such that (a) not all the s_i's are interpreted; (b) the only s_i's not interpreted are $\{I\}$ ($\{I\}$, of course, is the function which yields c_A, when applied to a context c). Where M is a meaning, we denote the set of i-reduced meanings of M, relative to c, by $M^{i,c}$. A member M_1 of $M^{i,c}$ is said to attribute a one-place property P just in case, for any context c' and world w

$$w \in M_1(c') \text{ iff } w \in P(c'_A).$$

When an $M_1 \in M^{i,c}$ and property P are so related, we write: $P \in [M^{i,c}]$. We can now say that the agent of a context c self-attributes the property P precisely if there is an M in V_c such that $P \in [M^{i,c}]$.

To define truth and denotation in an interpretation (reference to which is continually suppressed), we proceed as follows. The denotation of a term α, relative to a context c, assignment (member of U^V) f, and world w (write: $|\alpha|_{cfw}$) is defined: $f(\alpha)$, if $\alpha \in V$; c_A, if $\alpha = I$; C_{T_i}, if α is t_i; C_{Y_i}, if α is y_i. We begin as follows the definition of the notion ϕ, *taken relative to c and f, is true at w* (write: $cf[\phi]w$) as follows:

1. $cf[\Pi^n\alpha_1 \ldots \alpha_n]w$ iff $w \in [V(\Pi^n)(c)](\langle |\alpha_1|_{cfw}, \ldots, |\alpha_n|_{cfw}\rangle)$

2. $cf[(\phi) \wedge (\Psi)]w$ iff cf $[\phi]w$ and $cf[\Psi]w$.

And so on, for the other truth functors.

3. $cf[\exists\alpha(\phi)]w$ iff $\exists u(u \in U$ and $cf^{\alpha}_{u}[\phi]w)$.

Analogously for $\forall\alpha(\phi)$.

4. $cf[\alpha B^r(\phi)]w$ iff $\exists c'(c'_A = |\alpha|_{cfw}$ & $c'_w = w$ & $\exists m(m \in V(c')$ & $m(c') = \{w' | cf[\phi]w'\})$),

where $m(c')$ here is the proposition yielded by m in c', defined as above.

The intuitive content of clause (4) is this. $\alpha B^r(\phi)$, taken relative to c and f, is true exactly if: there is a meaning m such that α's denotatum accepts it (formally: $m \in V(c')$, c' the context of α's denotatum), and m yields, relative to c', that proposition expressed by ϕ relative to c. Note that this clause has the result (given that a person believes a proposition p if he accepts a meaning which yields p relative to his context) that $\alpha B^r(\phi)$ is true iff what α denotes believes the proposition expressed by ϕ.

Let $\hat{\alpha}(\phi)$ be a proper abstract. We say that P is the implied property of $\hat{\alpha}(\phi)$, taken relative to c and f, if and only if P is the one-place property such that, for all u and w,

$$w \in P(u) \quad \text{iff} \quad cf^{\alpha}_{u}[\phi]w.$$

We use $\overline{\hat{\alpha}(\phi)}^{cf}$ to denote the implied property of $\hat{\alpha}(\phi)$, taken relative to c and f. We may complete our definition of truth by saying that a *de se* ascription $\alpha B^s \hat{\beta}(\phi)$, taken relative to c and f, is true at w precisely if: α's denotatum believes a proposition under a meaning which has, as one of its *i*-reduced meanings, one which attributes $\overline{\hat{\beta}(\phi)}^{cf}$—that is, just in case α's denotatum self-attributes $\overline{\hat{\beta}(\phi)}^{cf}$. Formally, we have

5. $cf[\alpha B^s \hat{\beta}(\phi)]w$ iff $\exists c'(c'_A = |\alpha|_{cfw}$ & $c'_w = w$ & $\exists m(m \in V(c')$ & $\overline{\hat{\beta}(\phi)}^{cf} \in [M^{\iota,c}])$.

These semantics adequately capture the view of the truth conditions of *de se* ascriptions discussed at the beginning of this section. In particular, they have the consequence that a *de se* ascription implies (what we will presently define as) its corresponding *de re* ascription, although the converse implication does not hold. Thus, something of the form $\alpha B^s \hat{x}(\phi)$ involves an ascription of *belief*: the ascription is true only if α's denotatum believes the proposition ϕ expresses, when the denotatum of α is assigned to x.

We define the *de re* ascription corresponding to a *de se* ascription $\Psi = \alpha B^s \hat{x}(\phi)$ as follows. Let v be the least (i.e. with smallest subscript) variable not occurring in $\alpha B^s \hat{x}(\phi)$. The *de re* ascription corresponding to Ψ is then

$$\exists v(v = \alpha \wedge vB^r(\phi')),$$

where ϕ' is ϕ with all free occurrences of x replaced by v. (We of course understand the expression $\hat{\alpha}$ to bind free occurrences of α within its scope.) Thus, for example, corresponding to

$$IB^s\hat{x}_1(x_1B^s\hat{x}_1(Fx_1))$$

is

$$\exists x_2(x_2 = I \wedge x_2B^r(x_2B^s\hat{x}_1(Fx_1))).$$

It follows fairly directly from the above definitions that whenever a *de se* ascription, taken relative to c and f, is true at w, then so is its corresponding *de re* ascription. Of course, the converse does not hold. For example, if $V(c')$ consists solely of the meaning of 'Ft_5', 't_5'denotes c'_A relative to c',

$$\exists x_1(x_1 = I \wedge x_1B^r(Fx_1))$$

will be true, relative to c' and an assignment f, at c_w, but

$$IB^s\hat{x}(Fx)$$

will not.

There is a sense in which the semantics allows us to dispense with 'B^r' and make do with only 'B^s' as a belief predicate. For we can define 'B^r' using a schema along the lines of

$$\alpha B^r(\phi) = df \quad \alpha B^s\hat{\beta}(\beta = \beta \wedge \phi)$$

With some minor tinkering, this would be an adequate definition. (The tinkering required is this: As it stands, it is not the case that

$$\alpha B^r(\phi)$$

and

$$\alpha B^s\hat{\beta}(\beta = \beta \wedge \phi)$$

always agree in truth value, since (speaking very loosely) the latter's truth requires that the believer believe under the meaning of $\ulcorner I = I \wedge \phi \urcorner$, while the former requires simply belief under the meaning of ϕ. Now, although these meanings are identical when *conceived as functions* from contexts to propositions, they are not identical when conceived, as in our semantical system, as ordered n-tuples of the meanings of constituent expressions. Thus, to implement the above definition, we would need to impose a requirement on the function V in our models to the effect that $\{I = I \wedge \phi\} \in V(c)$, if $\{\phi\} \in V(c)$.)

However, such a definition has little, philosophically, to recommend it. The possibility of such a definition does not show that, in our regimentation, belief *de dicto* and *de re* are kinds of, or are reducible to, belief *de se*. (What it shows, I think, is that our system is committed to the thesis that anyone who believes a proposition *p* believes that he is himself and *p*, and the converse.) And it is certainly not the case that such a definition is what authors like Lewis and Chisholm have in mind when they suggest that belief *de re* is a kind of belief *de se*.

To take Lewis as an example: his view is that for someone *x* to believe *de re* of *u* that she is *F* is for *x* to self-ascribe the property *bearing R to one and only one thing, a thing that is F*, where *R* is a 'suitable' relation and *x* indeed bears *R* to *u* and *u* alone. On such a view, *de re* belief is not to be represented via quantification into the belief context (as we have represented it), nor will someone with such a view be sympathetic with our treatment of belief ascriptions involving demonstratives other than "*I*" (which is, in part, designed to represent such ascriptions as ascriptions of belief in propositions 'singular' with respect to the referents of the demonstratives). What is critical to regimenting Lewis's view is not eliminating 'B^r' in favour of 'B^s' (although that is involved), but giving a procedure for representing ascriptions, which appear to involve quantifying in, as not involving it.

Thus, we will preserve the operator "B^r", devoting the next section to a discussion of its semantics.

III

Our approach to *de se* ascriptions of belief is consistent with the view that the contents of *de dicto*, *de re*, and *de se* beliefs are all of the same category: they are all propositions. On the view just formalized, a (use of a) *de se* ascription of belief ascribes belief in a proposition. That is, for any such use *u*, there is a proposition, *p* such that *u* is true only if whomever belief is ascribed to, by *u*, believes *p*. But, on the approach we have suggested, that is not all such an ascription does; it also tells us something about the way (it implies that there is a particular way) in which belief is held.

Why should this be true only of *de se* ascriptions? Why, indeed: I believe that this is also true of *de re* ascriptions of belief.[15] I will argue in this section that there are pairs of *de re* ascriptions which ascribe to a person belief in the same proposition (given the theory of direct reference), but diverge in truth value. I will then discuss how a generalization of the semantics developed above can help the advocate of direct reference to account for this.

Consider *A*—a man stipulated to be intelligent, rational, a competent speaker of English, etc.—who both sees a woman, across the street, in a phone booth, and is speaking to a woman through a phone. He does not realize that the woman to whom he is speaking—*B*, to give her a name—is the woman he sees. He perceives her to be in some danger—a runaway steamroller, say, is bearing down upon her phone booth. *A* waves at the woman; he says nothing into the phone.

If *A* stopped and quizzed himself concerning what he believes, he might well say

(1) I believe that I can inform you of her danger via the telephone.

(It is understood here, and in the sequel, that uses of 'she' are accompanied by demonstrations of the woman across the street; uses of 'you' are addressed to the woman through the telephone.) *A* would deny the truth of an utterance by himself of

(2) I believe that I can inform her of her danger via the telephone.

The embedded sentences in (1) and (2) differ only with respect to demonstratives co-referential in the context. Hence (since the embedded sentences do not themselves contain any epistemological operators), if we accept the view that demonstratives are directly referential, we must say the embedded sentences express, relative to the contex, the same proposition. Thus, (1) and (2), taken relative to the context, ascribe to *A* belief in the same proposition.

Surely, however, (1) and (2) diverge in truth value here, (1) being true and (2) being false. One can muster convincing evidence for both these claims. To argue for the truth of (1), for example, we may first note that *A* surely knows what proposition he expresses when uttering its embedded sentence. For he knows the meaning of the sentence, he is perceiving the referents of the demonstratives therein, and may be said to know of each demonstrative that it denotes the thing perceived. (To forestall one sort of objection, suppose *B* to be speaking into the phone throughout the example.) Furthermore, the embedded sentence in (1) certainly seems to express something that *A* believes, namely 'I can tell *this* of the danger of *that* via the phone'. Given all of this, and the fact that *A* would, sincerely and after reflection, attest to the truth of (1), it seems that we ought to allow that (1) is true.

It does not follow, however, that (2) is true; indeed, it would seem that (2) is certainly not true. One argument one could advance in favour of this claim is this: (2) is true only if *A* believes that there is someone in danger with whom he can converse via the phone. As the case is set up, there is every reason to think that *A* does not have this belief. Hence, there is every reason to think that (2) is not true.

Presently, I will discuss what an advocate of the view that demonstratives are directly referential ought to say regarding cases like the one I have just presented. Before doing so, I shall digress in order to consider a case which, superficially, appears similar to one I have just outlined.

Consider again the situation of *A* and *B*. If *A* stopped and quizzed himself concerning what he believes, he might well sincerely utter

(3) I believe that she is in danger

but not

(4) I believe that you are in danger.

Many people, I think, suppose that here, again, we have a case in which sentences which ascribe belief to *A* in the same proposition (given that demonstratives are directly referential) clearly diverge in truth value, (3) being true and (4) being false.

It is clear that if we accept the thesis of direct reference, we must say that the embedded sentences in (3) and (4) express, relative to *A*'s context, the same proposition. But the view—that (3) is true in the context and (4) is not—is, I believe, demonstrably false. In order to simplify the statement of the argument which shows that the truth of (4) follows from the truth of (3), allow me to assume that *A* is the unique man watching *B*. Then we may argue as follows:

Suppose that (3) is true, relative to *A*'s context. Then *B* can truly say that the man watching her—*A*, of course—believes that she is in danger. Thus, if *B* were to utter

(5) The man watching me believes that I am in danger

(even through the telephone) she would speak truly. But if *B*'s utterance of (5) through the telephone, heard by *A*, would be true, then *A* would speak truly, were he to utter, through the phone

(6) The man watching you believes that you are in danger.

Thus, (6) is true, taken relative to *A*'s context. But, of course,

(7) I am the man watching you

is true, relative to *A*'s context. But (4) is deducible from (6) and (7). Hence,[16] (4) is true, relative to *A*'s context.

Note that a similar argument cannot be used to show that from the claim that (1) is true in *A*'s context, it follows that (2) is true. Consider how we might attempt to construct such an argument. We would have to argue that if *A* can truly utter (1), then *B* can truly say that the man watching her believes that he can inform her of her danger via the phone. That is, we would have to claim that if *A* can truly utter (1), then *B* can truly utter

(8) The man watching me believes that he can inform me of my danger via the phone.

Here, I think, the new argument goes awry. For it follows, from the claims that *B* can truly utter (8) and that *A* is the man watching *B*, that *A* believes that there is someone in danger who is such that he can tell her of her danger via the phone. But, as the case is set up, this is not so. Hence, there is no reason to think that an utterance of (8) by *B* would be true. (I shall discuss this further below.)

Let us now return to the original case. It is clear what we shall say about this case, if we accept the view of belief above labelled the triadic view. We shall say that *A* believes the proposition—that *B* can be informed of her danger via the phone—under the meaning of the embedded sentence of

(1) I believe that I can inform you of her danger via the telephone

but not under the meaning of the embedded sentence of

(2) I believe that I can inform her of her danger via the telephone.

This analysis should not be terribly puzzling, even given that *A* understands both sentences and knows of each, and the proposition it expresses, that the former expresses the latter. For, as *A* does not know that his uses of 'she' and 'you' are co-referential, he can hardly be expected to know that the embedded sentences express the same proposition.

Compare, now, the position of *A* with that of a person *X*, who is in the same situation as *A*, but who knows that the woman he sees is the woman to whom he is speaking. *X* will hold a belief about *B* under both the meanings mentioned above. He will also differ from *A* in the following way: there will be a woman whom *X* believes to have the property *being such that she can be informed of her danger via the phone*. It seems that we cannot explain this difference between *A* and *X* in terms of proposition believed, since both of them believe the proposition that *B* can be informed of her danger via the phone. In order to explain the difference, we must appeal to *how A* and *X* hold their beliefs. It would seem that to believe the proposition expressed (relative to a context *c*) by a sentence in which demonstratives occur is to have a *de re* belief with respect to the objects denoted, in *c*, by the demonstratives in the sentence. If one has a *de re* belief with respect to an object, then one may be said to *attribute* certain properties to the object. However, it does not follow, from the fact that *x* and *y* each believe the proposition *p* expressed in *c* by a sentence *S*(*d*), *d* a demonstrative occurring in *S* and denoting *u* in *c*, that every property which *x* attributes to *u*, in virtue of his believing *p*, is one which *y* attributes to *u*, in virtue of this belief. For which properties one attributes to an object is determined by the meaning under which one's belief is held: *X*, for example, who believes the proposition, that he can inform *B* of her danger via the phone, under the meaning of 'I can inform her of her danger via the phone', will attribute to *B* the property *being a thing that can be informed of its danger via the phone*: *A*, who does not believe the proposition under the meaning just mentioned, will not attribute this property to *B*.

If this much be accepted, we have the basis of an answer to the question: how can

(1) I believe that I can inform you of her danger via the telephone

and

(2) I believe that I can inform her of her danger via the telephone.

diverge in truth value in a context in which their embedded sentences express the same proposition? For we may say: an ascription of belief ⌜*a* believes that *S*⌝, *S* a sentence in which demonstratives occur, not only implies that the proposition expressed by *S* is believed, but that certain properties are attributed to the referents of the demonstratives in *S*. What properties the ascription implies are attributed depends, in turn, upon the meaning of *S*. In the case in question, ascription (2) implies that a property (that associated with a use, in this context, of 'I can inform *x* of *x*'s danger by phone') is attributed which (1) does not imply is attributed. Hence, (1) may true by while (2) is not.

Let us consider how we might give a systematic development of this proposal. In order to simplify matters, we will do this for a language with only a *de re* belief operator; it will be obvious how the treatment would be generalized to a language including a *de se* operator such as that discussed in Section II.

We assume, then, that our language has the same primitive vocabulary as the language of Section II, minus the B^s operator and the abstraction operator; the formation rules are identical to those of Section II, save the omission of the clause of the *de se* operator. We preserve the definitions of interpretation, denotation, and the clauses of the truth definition for atomic, truth functional, and quantified sentences. We now need to characterize, in terms of the formal structure, two things: when an individual, in believing a proposition under a meaning, attributes a property, and when a belief ascription, taken relative to a context, implies the attribution of a property.

Let $m = \langle\langle s_a, \ldots, s_n\rangle, M^n\rangle$ be a meaning. The intuitive answer to the question—when does the agent of a context *c* attribute a property *P*, in virtue of believing under *m*?—is as follows. Consider, first of all, what one 'gets' if one (*a*) replaces M^n with $M^n(c)$ (i.e. replaces the meaning M^n with the property which is its value in (*c*); (*b*) replaces each s_i either with its value in *c* or with a variable; (*c*) does not replace distinct s_i's with the same variable. Call such entities the *proto-properties* associated with *m* in *c*.

(For example, proto-properties associated with

$$m_1 = \langle\langle\{t_1\}, \{y_1\}\rangle, \{F^2_1\}\rangle$$

—which could be identified with the meaning of "$F^2_1 t_1 y_1$"—in a context in which "t_1" denotes u, "y_1" denotes u', and "F_1^2" denotes P are

(i) $\langle\langle u, x\rangle, P\rangle$

(ii) $\langle\langle x, u'\rangle, P\rangle$

(iii) $\langle\langle x, x'\rangle, P\rangle$.

Proto-properties associated with

$$M_2 = \langle\langle\{t_1\}, \{t_1\}\rangle, \{F_1^2\}\rangle$$

in such a context are all of the above and

(iv) $\langle\langle x, x\rangle, P\rangle$.)

To each proto-property there corresponds, in a rather obvious way, a property. For example: to (ii) corresponds the one-place property P^1 such that $w \in P^1(u_1)$ iff $w \in P(\langle u_1, u'\rangle)$; to (iii) corresponds the two-place property P^2 such that $w \in P^2(\langle u_1, u_2\rangle)$ iff $w \in P(\langle u_1, u_2\rangle)$; to (iv) corresponds the one-place property P^3 such that $w \in P^3(u_1)$ iff $w \in P(\langle u_1, u_1\rangle)$.

We can now answer our initial question thus: an agent attributes a property P, in virtue of holding a belief under a meaning m iff P corresponds to one of the proto-properties associated with m relative to the agent's context. We shall write

$$P \quad P \in P(m, c)$$

for the agent of c attributes P, in virtue of holding a belief under m.

A fully rigorous characterization of the above would dispense with the notion of a variable in the construction of proto-properties. It is easy enough to give such a characterization; we henceforth assume that the predicate $P(m, c)$ has been so defined in terms of our model structure. We now need a way to get from a sentence (taken relative to a context and an assignment) used to ascribe belief to the set of properties it implies the believer attributes. One way of doing this is as follows. Consider a sentence ϕ; let $\alpha_1, \ldots, \alpha_n$ be a complete enumeration of those demonstratives and variables (which occur freely) in ϕ. Let $v_1, \ldots, v_n$ be variables which do not occur in ϕ. We say that Ψ is a frame of ϕ just in case ϕ is the result of replacing one or more of the α_i's with v_i's, subject to the restriction that distinct α_i's are replaced with distinct v_i's.

Thus, for example, consider the sentences

(i) $F^2_2 t_1 y_1$

(ii) $F^2_2 t_1 t_1$.

Frames of (i) are: $F^2_2 t_1 x_1$, $F^2_2 x_1 y_1$, $F^2_2 x_1 x_2$; frames of (ii) are the above and $F^2_2 x_1 x_1$. Note that this last is not a frame of (i).

We say that a sentence ϕ implies the attribution of the property P^n, relative to c and f, just in case there is a frame Ψ of ϕ, obtained by substituting the n distinct variables $v_1, \ldots, v_n$ for terms in ϕ and, for every w and $u_1, u_2, \ldots, u_n$:

$$cf^{u_1 u_2 \ldots u_n}_{v_1 v_2 \ldots v_n}[\Psi]_w \quad \text{iff} \quad w \in P^n(\langle u_1, u_2, \ldots, u_n \rangle).$$

We define the attribution class of a sentence ϕ, relative to c and f, as the set of those properties such that ϕ implies their attribution, relative to c and f; we denote this class with $A(\phi, c, f)$.

We now define truth for *de re* ascriptions of belief:

$$cf[\alpha B^r \phi]w \quad \text{iff} \quad \exists c'(|\alpha|_{cfw} = c'_A \ \& \ c'_W = w \ \& \ \exists m(m \in V'_c \ \& \ m(c') = \{w' \mid cf[\phi]w'\} \ \& \ (h)(h \in A(\phi, c, f) \rightarrow h \in P(m, c')))),$$

where $m(c')$ is the proposition expressed by m relative to c'. Verbally, these truth conditions amount to this: $\alpha B^r \phi$, relative to c and f, is true exactly if there is a meaning m such that (i) $|\alpha|_{cfw}$ believes a proposition under m; (ii) m yields, relative to $|\alpha|_{cfw}$'s context, whatever ϕ expresses, relative to c and f, and (iii) whatever properties ϕ implies are attributed are such that belief under m requires their attribution.

It is easy to show that, given this semantics, representatives of sentences (1) and (2) can diverge in truth value relative to a context in which their embedded sentences express the same proposition.[17] On the other hand, the semantics validates the claim, for which we argued above, that in any context in which the uses of 'she' and 'you' in

(3) I believe that she is in danger

and

(4) I believe that you are in danger

are co-referential, the truth of (4) is implied by the truth of (3).

It is, perhaps, worth discussing sentences (3) and (4) again.

Many people, even after a rehearsal of the argument given above—that (4) is implied by (3)—are still uncomfortable with the claim that both (3) and (4) are true. A virtue of the semantics just presented, I think, is that it can be used to motivate an explanation of why the intuition, that (3) and (4) diverge in truth value, is so persistent.

Take a finite set of sentences and conjoin them; form what we called a frame of the result. (For example, if you start with {that_2 is sad, you_3 will make that_4 happy if that_2 helps you_3}, you will end up with something along the lines of 'x_2 is sad $\wedge$ x_3 will make x_4 happy if x_2 helps x_3'.) Call the property associated with such a sentence a *picture*; if all the members of the initial set are sentences, the meanings of which are accepted by an agent *u*, say that the resulting property is a picture *held by u*.[18]

The intuition motivating our semantical account is that an ascription is true provided it ascribes belief in a proposition which is believed and the ascription does not imply anything false about what pictures are held by the believer. Since sentence (4), as used by *A*, does not *when taken by itself* imply anything false about what pictures *A* holds, (4) so taken is true, since *A* believes *B* to be in danger.

Note, now, that a set of belief ascriptions may (conventionally) imply things about the pictures a believer holds that the conjunction of the members of the set does not (strictly) imply.[19] For example, the use of the ascription '*A* believes that you_1 are unhappy because she_2 spurned you_1', in a context in which the ascription '*A* believes that she_2 loves a Greek' has been used (and no one has disputed the truth of the latter ascription), will imply that *A* holds the picture associated with '*y* loves a Greek and *x* is unhappy because *y* spurned *x*'. Both ascriptions can be true, even if *A* does not hold this picture; however, their joint use, in such a case, would be very misleading.

In general, we tend to avoid using an ascription ⌜α believes that ϕ⌝, if an ascription ⌜α believes that Ψ⌝ is assumed by all the parties to the conversation to be true (and we know this) and we think that the person to whom belief is being ascribed does not hold pictures associated with frames of ⌜ϕ and Ψ⌝. Likewise, we will find an ascription ⌜α believes that ϕ⌝ bizarre or objectionable if it is assumed by those conversing that the ascription ⌜α believes that Ψ⌝ is true and we have good reason to think that the believer does not hold all the pictures associated with ⌜ϕ and Ψ⌝.

All of this, I believe, helps to explain why some find the assertion that *A*'s use of

(4) I believe that you are in danger

is true counter-intuitive, even after a rehearsal of the argument that *A*'s use of (4) cannot be false if his use of (3) is not. For as we have just seen, without qualification and explanation, the claim that (4) is true relative to *A*'s context is very misleading. For obviously, in the case under consideration

(9) I believe that I am talking to you

is true relative to *A*'s context. Thus, without further qualification, the claim that (4) is true implies that

(10) I believe that I am talking to someone who is in danger

is true, relative to *A*'s context. But, obviously, (10) is not thus true.

I close with some observations on the semantical theory suggested in this paper. According to this theory, ascriptions of belief are primarily, but not exclusively, vehicles for making reports about the content, as opposed to the manner, of belief. That ascriptions are primarily used to make reports about content ought not be surprising. For first of all, we are very often not in a position to say how a belief is held, although we know that it is held. (For example, one may know that Hank believes that Will spies, but not whether he accepts the meaning of 'that spies' or 'Will spies'.) Furthermore, it is often quite irrelevant to our purposes to specify how a proposition is believed.[20] Finally, we often cannot say how belief is held in any perspicuous way, even though we know. (Consider: Hank, Bernie, and Sally all believe that I am a spy.)

None the less, ascriptions of belief are, to a limited extent, used to report how belief is held. Indeed, in the semantics for ascriptions of belief suggested in this paper, the belief operator is construed as operating on sentential meanings, and not simply as an operator on the propositions which meanings, relative to a context, have as values. I have focused here upon relatively simple aspects of sentential meaning, in an attempt to make a case for the claim that, by construing the belief operator as an operator on meanings, as opposed to propositions, we can generate plausible solutions to semantical puzzles associated with the (quite plausible,

I believe) theory of direct reference. If the approach taken here strikes the reader as not without merit, he or she will, I hope, consider the question of how it is to be given the extensions and refinements it requires in order to yield a fully satisfactory theory.[21]

Notes

1. David Kaplan, *Demonstratives, Draft #2*, mimeograph UCLA: *Department of Philosophy*, 1977, p. 1. Henceforth, I will use 'demonstratives' as shorthand for 'demonstratives and indexicals'.
2. Thus, for example, Kaplan, in the section of *Demonstratives, Draft #2* entitled "Adding 'Says' " suggests truth conditions for (indirect discourse) ascriptions of belief which have the effect of making something of the form ⌜α believes that ϕ⌝ true exactly if α's referent believes (under any meaning whatsoever) the proposition the semantics assigns to ϕ. Elsewhere in *Demonstratives, Draft #2* Kaplan claims that all (non-quotational) operators of English are 'at most intensional'—i.e. they all can be construed as operating on (the formal representatives of) propositions.
3. An excellent discussion of what the thesis of direct reference does and does not imply can be found in Nathan Salmon, *Reference and Essence* (Princeton University Press, 1981).
4. I am adopting here some of the terminology and semantic assumptions of Kaplan's *Demonstratives, Draft #2* and 'On the Logic of Demonstratives', in Peter French *et al.* (eds.), *Contemporary Perspectives in the Philosophy of Language* (University of Minnesota Press, 1979); also in this volume.
5. I have argued in 'Tense, Propositions, and Meanings', *Philosophical Studies* 41 (1982), 337–51, that tense operators can be given an adequate semantical treatment only on the assumption that they operate on the meanings of, not simply on the propositions expressed by, sentences. Thus, questions about 'believes that' to one side, I think the possibility of there being an operator such as *O* is not at all unlikely.
6. See Kaplan, *Demonstratives, Draft #2*, and John Perry, 'The Problem of the Essential Indexical', *Nous* 13 (1979), 3–21 (also in this volume), and 'Belief and Acceptance', in Peter French *et al.* (eds) *Midwestern Studies in Philosophy*, vol. v (University of Minnesota Press, 1980).
7. This is closer to Kaplan's view than Perry's. On Perry's account, the second term in the relation is what Perry calls a *belief state*, which is a mental state individuated (in part, at least) in terms of the sentence types (or meanings thereof) which an agent in that state accepts, where *acceptance* is a technical term with a meaning related to (but probably not identical with) the meaning the term is accorded below in the text.

   I characterize the triadic theory as in the text because I find it easier to motivate the formalism of Sections II and III in terms of such a characterization.
8. I ought to say something here about what these meanings are, and how they differ; what needs to be made clear is what the meaning of terms like 'you' and 'she' is.

   I presume the following (and do not suggest that it is an original view; it is a version of Kaplan's own view). There are what we might call 'modes of demonstrating' things and 'modes of addressing' things. These modes are such that the same mode can be used in different contexts or several times in one context. It is only when 'she' is accompanied by a mode of demonstrating ('you' is accompanied by a mode of addressing) that it refers to an object. Furthermore, although 'she' plus mode m of demonstrating ('you' plus mode m' of addressing) may pick out different objects in different contexts, 'she' accompanied by one mode of demonstrating picks out the same object every time it is used in a context; analogously for 'you'.

The meaning (in Kaplan's sense of meaning as character) of 'she', then, is roughly this: 'she', accompanied by a mode of demonstrating, functions as a directly referential term; it denotes, relative to a context, what its accompanying mode of demonstrating demonstrates.

Thus, in giving formal representatives for sentences such as those mentioned in the text, what we really represent is the sentence type and aspects of the modes of demonstration or address. (For we wish to be able to assign the representatives of propositions to the formal representatives of sentences; the sentences being represented do not express propositions, on the view assumed here, unless accompanied by modes of demonstration or address.) We thus represent two occurrences of 'she' (of 'you') with the same term if and only if they are accompanied by the same mode of demonstration (or address).

These details will be germane to the view of *de re* belief ascriptions discussed in Section III.

9. For Kaplan's views, see n. 2. Perry has suggested in conversation that he accepts something along the general lines of the semantical view expressed in the sentence to which this is a note.
10. As a referee pointed out, it is misleading to single out (1) as 'the form' of *de se* ascriptions in English. This is, firstly, because sentences of the form ⌜believe that I am F⌝ seem, at least sometimes, to be used to ascribe *de se* belief and sometimes merely *de re* belief. Secondly, some sentences (e.g. 'I believe that Edwina will build a house near mine') seem to be used to report belief *de se* but are neither of the form of (1) nor such that they have a colloquial equivalent of the form of (1). (A further worry is whether or not (1) has a reading on which it is equivalent to (2); whatever the answer to this question, I do not think it will affect the points made in this section.)

    I will persist in speaking as if (1) gave the canonical form of *de se* ascriptions—a fiction which, I hope, is no more harmful in this context than the common fiction, in discussions of belief *de re*, of pretending that ⌜a believes, of b, that she is F⌝ is unambiguously *de re*, while ⌜a believes that b is F⌝ is unambiguously *de dicto*.
11. A sampler of such arguments is to be found in Roderick Chisholm's *The First Person* (University of Minnesota Press, 1981). It is not my purpose here to defend any particular argument as showing that the implication fails. Rather, I assume that it is *very* plausible that the implication does fail. Given this assumption, the question arises: How could an advocate of the view that demonstratives and indexicals are directly referential account for this?
12. To those familiar with views of *de se* belief advanced by Chisholm in *The First Person* and David Lewis in 'Attitudes *De Dicto* and *De Se*', *Philosophical Review* 87 (1979), 513–43, this will sound somewhat familiar. Chisholm introduces a primitive notion *x directly attributes property P to y* which,

according to Chisholm, is necessarily reflexive. Chisholm then says that to believe oneself to be F is to directly attribute F to oneself. Lewis suggests that we understand belief *de se* as the *self-ascription* of property.

There are several important differences between our approach and the approaches of Chisholm and Lewis. We do not hold that properties are the objects of *de se* belief, as do Lewis and Chisholm; we also hold that the objects of all beliefs are of uniform character, unlike Chisholm.

On Chisholm's view, it is somewhat mysterious as to why one can directly attribute properties only to oneself. Indeed, for Chisholm, there is no real correlate of direct attribution, relating distinct individuals and a property: Chisholm's indirect attribution (in terms of which Chisholm defines *de re* belief) is simply a complicated form of direct attribution.

On our view the reflexivity of self-attribution is not mysterious at all: It is reflexive because it involves meanings which contain $\{I\}$. Furthermore, we could define a perfectly analogous notion of indirect attribution, without invoking the notion of self-attribution, if we wished. Indeed, something like this is defined in Section III, below.

We have analogous differences with Lewis, who characterizes belief *de re* in 'Attitudes *De Dicto* and *De Se*' as a kind of belief *de se*. (For Lewis, as for us, the objects of belief are of uniform character; but, unlike us, he takes them to be all properties.)

It is worth noting that the formalization introduced in this section could be used, with some alterations, to regiment Lewis's view. (The major alterations would be to drop the 'B^r' operator introduced below, translating English sentences of the form of *a believes that S*, where S involves no reflexives, as: $aB^s\hat{\alpha}\,(\alpha = \alpha \wedge \phi)$. One would also be required, in a formalization of Lewis's view, to prohibit quantification into 'B^s', and to come up with a scheme to represent *de re* ascriptions. This is discussed at the end of Section II.) This should not hide the fact that there are fundamental differences in motivation between Lewis and ourselves. Beyond those mentioned above, we note that this essay and its formalism is intended to function in the defence of the thesis of direct reference, a thesis which—insofar as it is bound up with what Lewis and Kaplan call 'haecceitism'—is anathema to Lewis.

13. It is difficult to come up with natural sounding English sentences which unambiguously capture these readings. I believe that anyone who takes the notion of *de se* belief seriously will agree that the beliefs represented by (4) through (8) are different beliefs; if the beliefs *are* different, an adequate treatment of belief ascriptions *de se* and *de re* ought to be able to differentiate them, syntactically and semantically.
14. The semantics presented here is modelled upon that of Kaplan's Logic of Demonstratives; see *Demonstratives, Draft #2* and 'On the Logic of Demonstratives' for a detailed exposition.
15. As will become clear below, I consider any ascription of the form ⌜*a* believes that ϕ⌝, which is such that ϕ has explicit occurrences of demonstratives, a *de re* ascription of belief.
16. I assume a definition of validity such as that which Kaplan gives for his Logic of Demonstratives. (See 'On the Logic of Demonstratives'.) I also assume (what is true in that logic) that if A follows from B and B is true in context c, then A is true in c.

17. We can also show that the semantics validates certain forms of 'quantifying in'. Precisely, given our semantics, we have:

    If β is a member of D which occurs in ϕ, then if $cf[\alpha B^r(\phi)]w$, then $cf[\exists v(\alpha B^r(\phi[\beta/v]))w$, provided that β is free for v in ϕ.

    (If our semantics had allowed for the possibility that members of D failed to denote in some contexts, this rule would have to be weakened. For simplicity's sake, we have not allowed for this possibility.) That such a rule is sound justifies, in part, the claim that something of the form of $[\alpha B^r(\phi)]$ is a *de re* ascription, provided that ϕ contains a member of D.

    Note that not every 'way of quantifying in' is permitted by our semantics. In particular, from

    (i) $t_1 = t_2 \wedge IB^r(F^2t_1t_2)$

    the formula

    (ii) $\exists x_1 \exists x_2(x_1 = x_2 \wedge IB^r(F^2x_1x_2)$

    follows, but

    (iii) $\exists x_1(x_1 = x_1 \wedge IB^r(F^2x_1x_1))$

    does *not* follow. Given our reasons for adopting the treatment we have adopted, of course, one would not want (iii) to follow from (i).
18. Strictly speaking, of course, we can associate properties with open sentences possibly containing demonstratives only relative to a context. My ignoring that here does not effect the point.
19. I must stress that 'implies' is being used in two senses in this sentence. The first use of 'implies' is quite weak (certainly not the sort of implication which preserves truth). Roughly, the use I intend here is the sort present in (typical) uses of 'His saying that the movie was boring implies that he did not like it'.
20. Note, however, that it *is* very often important to us to get across that belief is held under a meaning involving $\{I\}$. One reason for this is that we seem to presuppose the truth of a psychological theory which predicts how people will behave when they so believe (and when they have certain desires, etc.). To effectively make use of such a theory in everyday affairs—in particular, to justify predictions of behavior *via* the theory—we need a way to say that a person believes in the relevant way. It is for reasons such as this that English has a *de se* belief operator like that discussed in Section II. That we have no very general need, as we do for beliefs held under meanings involving $\{I\}$, to say that someone holds a belief under the meaning of a sentence involving $\{that\}$ or $\{you\}$ explains, I think, the absence of belief operators in English which single out beliefs held under such meanings.
21. I am indebted to David Auerbach, Edmund Gettier, Richard Grandy, and Harold Levin for comments on my syntax and these semantics. An anonymous referee for this journal also made useful comments on an earlier draft, for which I thank him or her. Part of the work on this paper was done while I held NEH grant FX-28919; I am grateful for this support.

XI

DIRECT REFERENCE, PROPOSITIONAL ATTITUDES, AND SEMANTIC CONTENT*

SCOTT SOAMES

I

What do we want from a semantic theory? A plausible answer is that we want it to tell us what sentences say. More precisely, we want it to tell us what sentences say relative to various contexts of utterance. This leads to the view that the meaning of a sentence is a function from contexts of utterance to what is said by the sentence in those contexts. Call this the propositional attitude conception of semantics.

Another semantic picture that has enjoyed considerable popularity is the truth conditional conception. According to it, the job of a semantic theory is to tell us what the truth conditions of sentences are. On this view, the meaning of a sentence can be thought of as a function from contexts of utterance to truth conditions of the sentence as used in those contexts.

Suppose now that we put the propositional attitude and the truth conditional conceptions together. If we do this, it is virtually irresistible to conclude that what is said by a sentence in a context consists in its truth conditions relative to the context. But what are truth conditions?

One natural idea, embraced by the ruling semantic paradigm, is that the truth conditions of a sentence, relative to a context, are the metaphysically possible worlds in which the sentence, as used in the context, is true. Such truth conditions can be specified by a recursive characterization of truth relative to a context and a world. This characterization implicitly associates with each sentence a

Scott Soames, 'Direct Reference, Propositional Attitudes, and Semantic Content', *Philosophical Topics* (15) 1987, pp. 47–87.

function representing its meaning. The value of the function at any context as argument is the set of metaphysically possible worlds in which the sentence, as used in the context, is true. It is this that is identified with what is said by the sentence in the context, when the propositional attitude conception of semantics is combined with this version of the truth conditional conception.

This identification is, of course, highly problematic. The first difficulty one notices is that if S and S′ are necessarily equivalent relative to a context, then they are characterized as saying the same thing, relative to the context. However, it is highly counterintuitive to hold that all necessary truths say the same thing, that the conjunction of a sentence with any necessary consequence of it says the same thing as the sentence itself, and so on.

A plausible pragmatic principle extends this difficulty to the propositional attitudes of speakers:

(1) A sincere, reflective, competent speaker who assertively utters S in a context C says (or asserts), perhaps among other things, what S says in C.

This principle reflects an incipient relational analysis of the attitude of saying, or asserting—an analysis that sees it as a relation between speakers and things which serve as the semantic contents of sentences. Once this analysis is accepted, it is a short step to view propositional attitude reports in accord with (2) and (3):

(2) An individual i satisfies ⌜x says (asserts) that S⌝ relative to a context C iff i stands in a certain relation R to the semantic content of S in C.

(3) An individual i satisfies ⌜x v's that S⌝ (where v = 'believes', 'knows', 'proves', 'expects', etc.) relative to a context C iff i stands in a certain relation R′ to the semantic content of S in C.

But now our difficulties are surely unmanageable. Let us characterize distribution over conjunction and closure under necessary consequence as follows:

Distribution over Conjunction
If an individual i satisfies ⌜x v's that P&Q⌝ relative to C, then i satisfies ⌜x v's that P⌝ and ⌜x v's that Q⌝ relative to C. (For example, anyone who asserts that P&Q asserts that P and asserts that Q.)

Closure Under Necessary Consequence
If an individual i satisfies ⌜x v's that P⌝ relative to C, and if every possible world in which P is true relative to C is a possible world in which Q is true relative to C, then i satisfies ⌜x v's that Q⌝ relative to C. (For example, anyone who asserts that P asserts everything that necessarily follows from P.)

The second main difficulty with our combined truth conditional and propositional attitude conception of semantics is that it equates distribution of a propositional attitude verb over conjunction with closure of the attitude under necessary consequence. For if Q is a necessary consequence of P, then the set of metaphysically possible worlds in which ⌜P&Q⌝ is true is the same as the set of worlds in which P is true. Given the identification of truth conditions with semantic content, this means that their semantic contents are the same. But then, a relational semantics of propositional attitude reports together with distribution over conjunction will yield closure under necessary consequence.

The problem is that for many propositional attitude verbs distribution over conjunction is a fact whereas closure under necessary consequence is not. My four-year-old son Greg has said many things, and whenever he says that P&Q he says that P and he says that Q. However, there are lots of necessary consequences of things he has said that he has left unasserted, for example that $2^9 = 512$, that first order logic is complete but undecidable, and that stones are made up of molecules.

A third difficulty with our semantic conception takes this problem one step further. The same considerations that lead to the view that beliefs and assertions are closed under necessary consequence lead to the view that no one has ever believed or asserted anything that could not have been true (in any metaphysically possible world). Since every Q is a necessary consequence of an impossible P, anyone who believes or asserts what P expresses believes or asserts everything. And surely, no one ever has, or could have, done that.

The semantic assumptions that lead to these difficulties can be summarized as follows:

(A1*a*) The semantic content of a sentence (relative to a context) is the collection of circumstances supporting its truth (as used in the context).

(A1*b*) The collection of circumstances supporting the truth of a sentence (as it is used in a context) = the set of metaphysically possible worlds in which it is true (relative to the context).

(A2) Propositional attitude sentences report relations to the semantic contents of their complements—i.e. an individual i satisfies ⌜x v's that S⌝ (relative to a context C) iff i bears R to the semantic content of S (relative to C).

(A3) Many propositional attitude verbs, including 'say', 'assert', 'believe', 'know', and 'prove' distribute over conjunction.

Since these assumptions lead to unacceptable results, one or more of them must be rejected.

The crucial assumptions are (A1) and (A2), which, in turn, are direct descendants of the two conceptions of semantics mentioned earlier. (A1*a* and *b*) represent the truth conditional conception, with metaphysically possible worlds taken as truth conditions. (A2) represents the propositional attitude conception, with the relational analysis of 'say' and 'assert' extended to propositional attitude reports generally. The need to give up one or the other of these assumptions makes it necessary to rethink the fundamental issues underlying these semantic conceptions.

I will focus on the truth conditional conception. Much of the support it has enjoyed comes from the familiarity of the possible worlds machinery plus the fact that the semantic content of a sentence (relative to a context) should determine the possible worlds in which it is true. However, there is a big difference between admitting that semantic content determines such truth conditions and claiming that it should be identified with them. What we need is some conception of semantics in which the content of a sentence determines, but is not determined by, the metaphysically possible worlds in which it is true.

There are two main ways in which such a conception might be developed. One way is to retain the basic assumption (A1*a*) of the truth conditional conception, while rejecting the characterization of truth conditions, or truth-supporting circumstances, as metaphysically possible worlds. The idea is to try and find some more finely grained circumstances that will distinguish among sentences true in the same worlds. The second way in which an appropriate semantic account might be developed is to give up (A1*a*) thereby abandoning the fundamental tenet of the truth conditional conception. In its place, one might substitute a conception of semantic contents as complex objects that encode much of the structure of the sentences that express them, and that determine sets of truth-supporting circumstances, without being identified with them.

In what follows, I shall argue for the second approach. The heart of my argument involves the interaction of propositional attitudes with the phenomenon of direct reference. Let us say that a singular term is directly referential iff its semantic content relative to a context (and assignment of values to variables) is its referent relative to the context (and assignment). Variables are the paradigm examples of such terms. In recent years, a number of arguments have been given for treating names and indexicals as directly referential as well. Later, I will show how this view can be defended against certain objections based on the behavior of such terms in propositional attitude ascriptions. To begin with, however, I wish to note the destructive consequences it has when added as a fourth assumption to (A1)–(A3):

(A4) Names, indexicals, and variables are directly referential.

This expanded set of assumptions has a number of clearly unacceptable consequences. Suppose, for example, that Mary assertively utters (4*a*), while pointing at me. On the assumptions we are considering, she cannot correctly be reported to believe, or to have said, that I am David Kaplan:[1]

(4*a*) He is David Kaplan (said pointing at Scott).

(4*b*) Mary says (believes) that he (Scott) is David Kaplan.

The reason for this is that the semantic content of the complement sentence, relative to the context, is taken to be the set of metaphysically possible worlds in which two distinct objects are absolutely identical with one another—that is, the empty set. But then the third difficulty noted above—the impossibility of saying or believing the impossible—comes into play, ruling out the possibility that Mary said or believed what she seemed to say and believe. The same problem arises in a variety of cases, including those in (5):

(5*a*) John says (believes) that Ruth Marcus is Ruth Barcan's sister.

(5*b*) Martin says (believes) that this table is made up of atomic particles with properties, P, Q, and R (where it is later discovered that nothing made of such particles could be a table).

The significance of these difficulties is not that they mar an otherwise unproblematic account of the attitudes. As we have seen, the conjunction of (A1)–(A3) is problematic in its own right. Nevertheless, the difficulties arising from the addition of (A4) are special.

I shall argue that these difficulties are intractable for theories that identify semantic contents of sentences with sets of truth-supporting circumstances. Although many of the problems encountered in standard, truth-theoretic accounts of the attitudes can be avoided by substituting fine-grained circumstances for metaphysically possible worlds, those posed by names and indexicals cannot. Not only do these problems resist such treatment, they remain even when assumptions (A2), (A3), and (A4) are weakened substantially. In effect, directly referential singular terms can be used to show that semantic contents of sentences (relative to contexts) cannot be sets of truth-supporting circumstances, no matter how fine-grained.

The reason for this is that such terms require the introduction of structure into semantic contents. After establishing this, I shall consider two different ways in which such structure might be constructed—one based on a modified version of the truth-theoretic approach, the other based on the introduction of structured, Russellian propositions. Although considerations involving directly referential singular terms are insufficient to decide between these alternatives, I shall argue that additional factors favour the Russellian approach. Thus, the end result is an argument for an expanded conception of semantics that includes Russellian propositions as semantic contents of sentences, over and above standard, truth-theoretic intensions and extensions.

II

Let us begin with the strategy of substituting fine-grained truth-supporting circumstances for metaphysically possible worlds. These circumstances can be thought of as arising from the relaxation of certain constraints that hold for such worlds. Taking a cue from Carnap's notion of a state description, we can describe these constraints in terms of their role in constructing a semantics for a language L.

Let D be the set of individuals L is used to talk about, and B be the set of properties expressed by simple predicates of L plus their complements.[2] Let us say that a C-description is a set each of whose members consists of an n-place property plus an n-tuple of objects drawn from D (for variable n). A C-description X is complete iff it contains a complete assignment of objects to properties—i.e. iff for every n-place property P in B, and every o_1, . . ., o_n in D, either $\langle P, o_1, \ldots, o_n \rangle$ is a member of X or $\langle [-P], o_1, \ldots, o_n \rangle$ is a member of X, where $[-P]$ is the complement of P. A C-description X is *consistent* iff no two of its members are negations of one another—i.e. iff for every n-place property P in B, $\langle P, o_1, \ldots, o_n \rangle$ is a member of X only if $\langle [-P], o_1, \ldots, o_n \rangle$ is not a member of X. A C-description is *metaphysically possible* only if it is metaphysically possible for the objects mentioned in the description to (jointly) instantiate the properties they are paired with in the description.

For present purposes, truth-supporting circumstances might either be identified with C-descriptions, or be taken to correspond to them. The classifications "complete", "consistent", and "metaphysically possible" can then be applied to circumstances.

Metaphysically possible worlds are truth-supporting circumstances that are metaphysically possible, complete, and consistent. Suppose the first of these constraints is relaxed, while retaining the second and third. This allows truth-supporting circumstances corresponding to every consistent and complete C-description. Thus, we allow metaphysically impossible circumstances in which Ruth Marcus is Ruth Barcan's sister, $2^9 \neq 512$, and I am identical with David Kaplan (' = ' being treated as a simple, non-logical predicate in the object language). In effect, we substitute what might be called "logically possible" worlds or circumstances for "metaphysically possible" worlds or circumstances.

However, the structure of the semantic theory remains the same as before. It continues to be a recursive characterization of truth relative to a context and circumstance, with the recursive clauses retaining their standard specifications. The semantic content of a sentence relative to a context is identified with the set of circumstances in which it is true. But since these circumstances are more finely grained than metaphysically possible worlds, we no longer have the results that metaphysically equivalent sentences have the same semantic content, that distribution of a propositional attitude verb over conjunction requires closure of the attitude under metaphysically necessary consequence, or that no one can believe or assert the metaphysically impossible. In this way, substitution of (A1*b*′) for (A1*b*) might be seen as alleviating the original difficulties with (A1)–(A4):

(A1*b*′) The collection of circumstances supporting the truth of a sentence (relative to a context) = the set of *logically possible worlds* in which it is true (relative to the context).

It does, of course, remain true on this view that logically equivalent sentences have the same semantic content, that distribution of a propositional attitude verb over conjunction requires closure of the attitude under logical consequence, and that no one can believe or assert the logically impossible. However, with another weakening of the constraints even these results can be avoided.

Suppose we give up the requirement that truth-supporting circumstances be complete. Instead we allow circumstances to correspond to (and, in effect, be exhausted by) any consistent C-description. Such circumstances are more like "logically possible facts" than "logically possible worlds". For example, one such circumstance may consist entirely of my being human.

The introduction of partial circumstances has import for certain logical constructions, most notably negation. In order to make semantic use of partiality, one must distinguish between it not being the case that in C an individual o has the basic property P, and it being the case that in C, o has the property of not being P. The latter is a truth-supporting circumstance for the negation of the atomic sentence that predicates P of o; the former is not. Full fledged negation, applied to sentences of arbitrary complexity, as well as related constructions like material implication, raise complications that we need not go into. However, other constructions are straightforward. For example, the recursive clauses governing conjunction, disjunction, and existential generalization are exactly those used in standard, truth-theoretic accounts.

The semantic content of a sentence relative to a context is, as usual, the set of circumstances supporting its truth, as used in the context. However, since circumstances are partial, the semantic contents of logically equivalent sentences are no longer identified. For example, the content of (6*a*) is not the same as the content of (6*b*), because the former includes "facts" that are, so to speak, silent about radioactivity:

(6*a*) Plymouth Rock is in Massachusetts.

(6*b*) Plymouth Rock is in Massachusetts & (Plymouth Rock is radioactive v Plymouth Rock is not radioactive).

This is significant, since, it might be argued, a person lacking the concept of radioactivity might believe that which is expressed by (6*a*) without believing that which is expressed by (6*b*). Certainly, it would seem that someone could assert the former without asserting the latter. One way of accounting for this within the framework of (A1–A4) is to substitute (A1*b*″) for (A1*b*′):

(A1*b*″) The collection of circumstances supporting the truth of a sentence (relative to a context) = the set of *logically possible facts* that would make it true (as used in the context).

This strategy is followed by Jon Barwise and John Perry in their book *Situations and Attitudes*. However, they take it one step further, allowing truth-supporting circumstances to be inconsistent, as well as incomplete and metaphysically impossible. If one ignores complications involving time, tense, and spatio-temporal location, one can take their "abstract situations" to be arbitrary C-descriptions.[3] Allowing these circumstances to be inconsistent, and substituting (A1*s*) for (A1*b*″), makes it possible to correctly characterize certain agents as believing and asserting contradictions —e.g. as believing and asserting that London is pretty and London is not pretty.

(A1*s*) The collection of circumstances supporting the truth of a sentence (relative to a context) = the set of *abstract situations* which would make it true (as used in the context).

Logically complex constructions are characterized along familiar truth-theoretic lines. For example, we have:

(7*a*) The semantic content of a conjunction (relative to a context) is the intersection of the semantic contents of the conjuncts (relative to the context).

(7*b*) The semantic content of a disjunction (relative to a context) is the union of the semantic contents of the disjuncts (relative to the context).

(7*c*) The semantic content of an existential generalization ⌜For some x: Fx⌝ (relative to a context) is the set of circumstances E such that for some object o in E, o "is F" in, or relative to, E (and the context).[4]

(7*d*) The semantic content of ⌜F[an x: Gx]⌝ (relative to a context) is the set of circumstances E such that for some object o in E, o "is G" and o "is F" in, or relative to, E (and the context).

(7*e*) The semantic content of ⌜F[the x: Gx]⌝ (relative to a context) is the set of circumstances E such that for exactly one object o in E, o "is G" in, or relative to, E (and the context); and, moreover, o "is F" in, or relative to, E (and the context).

The invariance of these principles across different choices of truth-supporting circumstances reflects the fact that no matter what one's conception of circumstances, the circumstances that make a conjunction true are those that make the conjuncts true; the circumstances that make a disjunction true are those that make either disjunct true; and so on. Indeed, we may take the principles in (7) to be partially constitutive of the view that the semantic content of a sentence consists in the circumstances that support its truth. As such, they may be regarded as corollaries of assumption (A1*a*).

There is, then, a whole range of possible theories within the standard, truth conditional framework that adopt the same basic approach to the problems posed by various kinds of propositional attitudes. The central idea is to relax the constraints on truth-supporting circumstances. This results in more finely grained semantic contents being attached, in the first instance, to atomic sentences. Logically complex constructions are given the usual recursive treatment, resulting in semantic contents for complex sentences along the lines of (7).

This approach can be seen as an attempt to save the truth conditional conception of semantic content, while countenancing direct reference and continuing to take semantic contents of sentences to be objects of propositional attitudes. Although not without plausibility, it is, I believe, fundamentally flawed. Its chief virtue is its recognition that if assumptions (A2), (A3), and (A4), plus an elementary principle of compositionality,[5] are to be retained, then semantic contents must be more fine-grained than sets of metaphysically possible worlds. Its chief error is its failure to recognize that if these assumptions are retained, then no conception of truth-supporting circumstances validating (7) can do the job, no matter how fine-grained.

III

A number of different arguments can be used to show this. For example, consider (8):

(8*a*) The ancients believed (asserted) that 'Hesperus' referred to Hesperus and 'Phosphorus' referred to Phosphorus.

(8*b*) The ancients believed (asserted) that 'Hesperus' referred to Hesperus and 'Phosphorus' referred to Hesperus (from (A2), (A4), and compositionality in the complement).

(8*c*) The ancients believed (asserted) that 'Hesperus' referred to Hesperus and 'Phosphorus' referred to Hesperus and for some x, 'Hesperus' referred to x and 'Phosphorus' referred to x (from (A1*a*) and (A2)).

(8*d*) The ancients believed (asserted) that for some x, 'Hesperus' referred to x and 'Phosphorus' referred to x (where the quantifier is inside the scope of the propositional attitude verb). (From (A3).)

Since (8*d*) is tantamount to the claim that the ancients believed and asserted that the terms 'Hesperus' and 'Phosphorus' were coreferential, it is false. Since (8*a*) can be regarded as true, at least one of the principles used in going from (*a*) to (*d*) must be rejected.

The first thing to note is that these principles do not include (A1*b*), (A1*s*), or any other specific characterization of truth-supporting circumstances. The only use made of truth-supporting circumstances was the appeal to (7*a*) and (7*c*) in the move from (*b*) to (*c*) in the argument. Since these principles are corollaries of (A1*a*), acceptance of the other assumptions in the argument requires rejection of the claim that the semantic content of a sentence (relative to a context) is the set of circumstances supporting its truth (as used in the context).

The same point can be made using definite descriptions instead of existential quantification. For example, consider (9):

(9*a*) y believes (asserts) that Hesperus = the x:Fx and Phosphorus = the x:Gx.

(9*b*) y believes (asserts) that Hesperus = the x:Fx and Hesperus = the x:Gx (from (A2), (A4), and compositionality in the complement).

(9*c*) y believes (asserts) that Hesperus = the x:Fx and Hesperus = the x:Gx and the x:Fx = the x:Gx (from (A1) and (A2)).

(9*d*) y believes (asserts) that the x:Fx = the x:Gx (where the descriptions are used attributively and are within the scope of the propositional attitude verb). (From A3.)

The move from (*b*) to (*c*) is justified if every circumstance supporting the truth of the complement of (*b*) supports the truth of the complement of (*c*). One gets this if circumstances are metaphysically possible worlds, since any world in which o is identical with o′ and o″ is a world in which o′ and o″ are identical.

However, there is no need to rest the case on special assumptions about circumstances, or identity. By recasting the example one can make use of the semantics (7*e*) for definite descriptions to construct an argument that applies to all the theories in section II. One simply starts with (9*a*′) instead of (9*a*):

(9*a*′) y believes (asserts) that Hesperus = the x:Fx and Phosphorus = the x:Gx and the x:Fx = the x:Fx and Hesperus = the x such that Hesperus = x.

It follows from (7*a*) that a circumstance E will support the truth of the complement of (9*a*′)) iff it supports the truth of each of its conjuncts. It follows from (7*e*) that E will support the truth of the final conjunct iff there is exactly one object o such that Hesperus = o in E. Since Hesperus is Phosphorus, this means that o must be both the unique F-er in E and the unique G-er in E. The third conjunct requires that o = o in E. This guarantees that E will be a member of the semantic content of the complement of (9*d*). Thus, (A2), (A3), (A4), and a principle of compositionality allow one to derive (9*d*) from (9*a*′), no matter how finely grained one takes truth-supporting circumstances to be.[6] Since (9*d*) may be false even when (9*a*′) is true, acceptance of (A2), (A3), (A4), and the compositionality principle requires rejection of (A1*a*).

A more startling illustration of this conclusion can be constructed using the examples in (10):

(10*a*) Mark Twain = Herman Melville and Samuel Clemens = Stephen Crane.

(10*b*) Mark Twain = the x such that Mark Twain = x.

(*a*) is an embarrassment to standard treatments of the attitudes (encompassing (A2)–(A4)) in which truth-supporting circumstances are taken to be metaphysically possible worlds. Since its semantic content in such systems is the empty set, everything is a semantic consequence of it. Thus, that which it expresses cannot be believed or asserted.

One of the virtues of systems that relax constraints on truth-supporting circumstances is that they avoid this embarrassment. In such systems the semantic content of (10*a*) is a non-empty set of circumstances in which three distinct individuals are identified. Although such circumstances are metaphysically impossible, they are regarded as semantically legitimate, and hence are available for the construction of semantic contents. Thus, it is perfectly possible, in a system like that of *Situations and Attitudes*, for a person to believe or assert that which is expressed by (10*a*).

Belief or assertion of that which is expressed by (10*b*) is unproblematic on any account. However, now consider their conjunction, (10*c*):

(10*c*) Mark Twain = Herman Melville and Samuel Clemens = Stephen Crane and Mark Twain = the x such that Mark Twain = x.

In order for a circumstance E to be a member of the semantic content of this sentence, E must be a member of the semantic content of each conjunct. In order for E to be a member of the semantic content of the first two conjuncts it must be the case that in E Mark Twain is identified with two distinct individuals. But now E cannot be a member of the semantic content of the third conjunct, since, by (7*e*), that conjunct requires that Mark Twain be identified with only one object. The semantic content of (10*c*) is, therefore, the empty set. Thus, the problems posed by (10*a*) for theories embracing the original (A1)–(A4) are reproduced by (10*c*) for theories that substitute finer grained truth-supporting circumstances for metaphysically possible worlds.[7]

Although this example is particularly graphic, the basic difficulty is extremely general. It is repeated in (11), where (*b*) is derived from (*a*) using the semantics (7*d*) for indefinite descriptions, and in (12), where a similar derivation uses material implication:[8]

(11*a*) x believes (asserts) that Mark Twain wrote the greatest

American novel and Samuel Clemens was an ignorant illiterate.

(11*b*) x believes (asserts) that an ignorant illiterate wrote the greatest American novel (where the indefinite description is attributive and inside the scope of the propositional attitude verb).

(12*a*) x believes (asserts) that Mark Twain is F and if Samuel Clemens is F then S (where F is any predicate and S is any sentence).

(12*b*) x believes (asserts) that S.

The difficulty common to all these cases is, I suggest, not due to special assumptions about particular constructions (existential quantification, definite descriptions, indefinite descriptions, conjunction, material implication, etc.) Rather, the general assumptions (A1*a*), (A2), (A3), and (A4) (plus compositionality in the complements of propositional attitude verbs) are jointly incompatible with facts about propositional attitudes and propositional attitude ascriptions. In short, we have established (13):

(13) If direct reference is legitimate and (some) propositional attitude verbs have a relational semantics ((A4) plus (A2)), then (assuming compositionality and distribution over conjunction) the semantic contents of sentences relative to contexts cannot be sets of truth-supporting circumstances (that validate 7).

This way of putting the matter is, of course, not neutral, since it suggests that the assumption to be rejected is (A1*a*). This suggestion can be supported by showing that the remaining assumptions are both stronger than needed to refute (A1*a*) and more plausible than they might initially appear.

IV

First consider (A4). The arguments in section III all involve proper names. Thus, one response to them might be to give up the claim that names are directly referential, thereby blocking substitution of coreferential names in propositional attitude ascriptions. It is important to note that this response is insufficient, since, in each case the problem can be recreated using other terms.

For example, so long as direct reference is retained for demonstratives, (A1*a*), (A2), (A3), and compositionality will allow one to derive the false (14*b*) from the potentially true (14)*a*):

(14*a*) The ancients believed (asserted) that their such-and-such utterance referred to this (pointing in the morning to Venus) and (speaking very slowly) their so-and-so utterance referred to that (pointing in the evening to Venus).

(14*b*) The ancients believed (asserted) that for some x, their such-and-such utterance referred to x and their so-and-so utterance referred to x.

The same point can be made using variables in place of names and indexicals:

(15*a*) There is a planet x which is seen in the morning sky and a planet y which is seen in the evening sky and the ancients believed that x was seen in the morning sky and y was seen in the evening sky.
(∃x: Px&Mx) (∃y: Py&Ey) (a believed that (Mx&Ey)).

(15*b*) The planet seen in the morning sky is the planet seen in the evening sky.
the x:(Px&Mx) = the y:(Py&Ey).

(15*a*) is true iff there is an assignment f which assigns a planet seen in the morning sky to 'x' and a planet seen in the evening sky to 'y' such that the open belief sentence is true with respect to f. From (15*b*) it follows that the referents of 'x' and 'y' with respect to f are identical. But now (A1)–(A4) can be applied as before to derive the false (15*c*–*d*) from the true (15*a*–*b*):[9]

(15*c*) There is a planet x and a planet y such that the ancients believed the following: that x was seen in the morning sky and y was seen in the evening sky and there was something which was (both) seen in the morning sky and seen in the evening sky.
(∃x: Px) (∃y: Py) (a believed that ((Mx&Ey) & ∃z(Mz&Ez))).

(15*d*) The ancients believed that there was something which was (both) seen in the morning sky and seen in the evening sky.
(a believed that ∃z(Mz&Ez)).

Thus, if direct reference is the source of the difficulty, it must be banned entirely—for names, indexicals, and variables. But this is implausible; the arguments for it are too strong, and there are too many cases (where the words of the speaker differ systematically from those of the agent of the attitude) in which it is instrumental in capturing clear semantic intuitions.

There is, however, another way in which one might try to block the problematic arguments. Each of them relies on assumptions—(A2), compositionality, and some version of direct reference—that jointly legitimate the substitution of coreferential terms in propositional attitude ascriptions. It might be thought that such substitution is the source of the problem. As against this, it is worth noting that the difficulty can be recreated without appealing to substitutivity, or the assumptions that give rise to it.

Instead of relying on semantic analyses of propositional attitude statements one can invoke principles underlying our practice of reporting propositional attitudes and ascribing them to individuals. Why, for example, do we ascribe to the ancients the belief and assertion that Hesperus was visible in the evening, while being reluctant (at least initially) to ascribe to them the belief and assertion that Phosphorus was visible in the evening? Probably because they assertively uttered sentences whose English translation is 'Hesperus is visible in the evening', but refused to assertively utter (and indeed dissented from) sentences whose English translation is 'Phosphorus is visible in the evening'. These examples suggest (if we focus on indexical-free sentences and ignore complications involving time and tense) the following principles of propositional attitude ascription:

(16*a*) If a competent speaker x of a language L sincerely and reflectively assents to (or assertively utters) an indexical-free sentence s of L, and if p is a proper English translation of s, then x satisfies ⌜y believes that p⌝. (Note that this covers the case in which L = English and s = p.)

(16*b*) If a sincere, reflective, and competent speaker x of a language L assertively utters an indexical-free sentence s of L, and if p is a proper English translation of s, then x satisfies ⌜y says (asserts) that p⌝.

These principles are, of course, modelled after Kripke's principles of (weak) disquotation and translation.[10] With them we can derive the conclusion that Kripke's bilingual speaker Pierre believes and asserts both that London is pretty and that

London is not pretty. The former follows from his sincere and reflective utterance of 'Londres est jolie', plus (16) and an elementry truth of translation. The latter follows from his equally sincere and reflective utterance of 'London is not pretty', plus either (16) alone, or (16) in conjunction with homophonic translation.

It seems to me that these ascriptions to Pierre are correct. It is, of course, striking that Pierre's beliefs and assertions should be contradictory without his having made any mistake in logic or reasoning. However, this just shows that in certain cases one may be in no position to determine the consistency of one's statements and beliefs.

The point is particularly obvious in the case of what is said or asserted. Imagine Pierre on the telephone talking to a friend in Paris. During the course of the conversation he assertively utters 'Londres est jolie'. After hanging up the phone he says 'London is not pretty' to a visitor who asks his opinion of the city he lives in. What has Pierre said? Clearly, he has said both that London is pretty (to his friend) and that London is not pretty (to the visitor).

Now consider a slight extension of the example. Suppose that there are a number of Frenchmen in London in the same linguistic and epistemic situation as Pierre. When together they converse with one another in French—standard French plus one addition. Since they are unaware that 'Londres' names the city they live in, they use the name 'London' for that purpose. One day Pierre assertively utters 'Londres est jolie et London n'est pas jolie'. I, an English speaker, am asked to report what he said. Since Pierre is competent in his own dialect, I can appeal to (16). Since his dialect is one in which both 'London' and 'Londres' are properly translated into English as 'London', I can report that he said (asserted) that London is pretty and London is not pretty. To avoid puzzling my audience, I will, of course, say more than this. However, the initial report is surely correct. In certain cases two words in one language do have the same translation into a second language (e.g. 'Peking' and 'Bejing' in English); and assertive utterances by normal, competent speakers can be reported in indirect discourse of the second language.[11]

This fact can be used to reconstruct the arguments of section III without appealing to direct reference, compositionality, or substitutivity at all. In the case at hand, we have used (16*b*) plus a truth of translation to establish (17):

(17) Pierre said (asserted) that London is pretty and London is not pretty.

To derive (18):

(18) Pierre said (asserted) that London is pretty and London is not pretty and for some x, x is pretty and x is not pretty.

we need only appeal to corollaries (7*a*) and (7*c*) of (A1*a*), plus the following weakened version of (A2):

(A2′) An individual i satisfies ⌜x v's that S⌝ (relative to a context C) iff i bears a certain relation R* to the pair consisting of the content of S (relative to C) and the character of S (i.e. the function from contexts to contents that represents the meaning of S).[12]

(19) follows from (18) by (A3):

(19) Pierre said (asserted) that for some x, x is pretty and x is not pretty.

But (19) is false—Pierre did not assert the proposition that something is both pretty and not pretty. Thus, we have another reductio of (A1*a*), this time from a considerably weakened set of premisses. Similar reductios can be constructed corresponding to each of the arguments in III.[13]

However, the premisses are still stronger than they need to be. Although (A3) is useful in deriving obviously false conclusions, it is not strictly necessary. (8*c*), (8*c*′), (9*c*), (9*c*′), (15*c*), and (18) are all false, and can be derived without (A3):[14]

(8*c*′) The ancients believed (asserted) that 'Hesperus' referred to Hesperus and 'Phosphorus' referred to Phosphorus and for some x, 'Hesperus' referred to x and 'Phosphorus' referred to x.

(9*c*′) y believes (asserts) that Hesperus = the x:Fx and Phosphorus = the x:Gx and the x:Fx = the x:Gx.

Even (A2), and its weakened counterpart (A2′), may give a misleading impression of strength. As presently formulated, they ignore one possible type of semantic information—to wit, information fixing the referent of a name as a matter of linguistic convention. I suspect that arabic numerals are names that carry such information.[15] Some might hold that 'Hesperus' and 'Phosphorus' are, too.[16] If they are, then the weakened principle (A2*) will block substitution of one for the other in propositional attitude ascriptions:[17]

(A2*) An individual i satisfies ⌜x v's that S⌝ (relative to a context C) iff i bears a certain relation R** to the triple consisting of the content of S (relative to C), the character of S, and an n-tuple of properties $\langle P_1, \ldots, P_n \rangle$, where P_i fixes, as a matter of linguistic convention, the referent of the ith name in S.

However, such a move will not block the reductio of (A1*a*). First, not all proper names have conventionally associated reference fixing properties. Second, as Kripke has shown, variants of the Pierre case can be constructed in which the names 'London' and 'Londres' *are* associated with the same properties (provided they are not "purely qualitative").[18] Finally, substitution of one term for another is not always required for refutations of (A1*a*). Suppose, for example, that 'Hesperus' and 'Phosphorus' share the same object as content and the same constant function from contexts to that object as character, but differ in reference fixing properties. Although (A2*) will then block the derivation of (8*b*) and (8*c*) from (8*a*), it will still allow the derivation of (8*c*′). (The same goes for (15).)[19]

Results like these suggest that the reductio of (A1*a*) cannot be blocked by any plausible weakening of the subsidiary premisses used in the original argument. It is true that those premisses jointly give rise to some surprising, and initially counterintuitive, results involving substitution in propositional attitude ascriptions. However, the reductio can be recreated (in a variety of ways) even when those results are avoided, or minimized.

A final illustration of this point is provided by the following example: Professor McX, looking through the open back door of the faculty lounge, sees Y walking down the hall and says to a visitor, "He (pointing to Y) is a professor in the department." A few seconds later Y passes by the front door, and McX says "He (pointing to Y again) is a graduate student in the department." Although McX does not realize that he has pointed twice to the same individual, Y, who has overheard the remarks, can correctly say, "McX said both that I am a professor in the department and that I am a graduate student in the department."

Developing the example further, we can have McX conjoin his remarks:

(20) Who is in the department? Let me see. He (pointing to Y as he passes the back door) is a professor in the department and (turning) he (pointing to Y as he passes the front door) is a graduate student in the department.

On the basis of McX's remark, Y says:

(21) McX said that I am a professor in the department and I am a graduate student in the department.

Y's assertion is unexceptionable. Unlike some other examples we have considered, this one does not require the creation of an unusual situation; it does not involve attributing conflicting statements (or beliefs) to an otherwise rational agent; nor does it raise the suspicion that adherence to otherwise plausible principles forces us to accept a counterintuitive result. Whatever semantic analysis of propositional attitude ascriptions turns out to be correct, Y's report is one that we want, pre-theoretically, to come out true.

This is not the case with (22) (where the quantifier is understood as being inside the scope of the propositional attitude verb):

(22*a*) Professor McX said (asserted) this: that there is at least one x such that x is a professor in the department and x is a graduate student in the department and I am a professor in the department and I am a graduate student in the department.

(22*b*) Professor McX said (asserted) that there is at least one x such that x is a professor in the department and x is a graduate student in the department.

These reports are clearly not true.[20]

If this is correct, then the problem for (A1*a*) is obvious. Corollaries (7*a*) and (7*c*) of that principle characterize the complements of (21) and (22*a*) as having the same content (with respect to the context). But then there will be no semantic value (content, character, or reference fixing properties) differentiating them. As a result, virtually any relational semantics of assertion-ascriptions (e.g. (A2), (A2′), (A2*)) will assign (21) and (22*a*) the same truth value. (A3) will then extend this error to (22*b*). Since these results are unacceptable, while relational treatments of assertion and other attitudes remain plausible, (A1*a*) should be rejected.

V

We have just seen that the impossibility result of section III can be reproduced using (A1*a*) together with sets of auxiliary premisses considerably weaker than the original (A2), (A3), (A4), and compositionality. This constitutes an important reason for taking that result to be a reductio of the assumption that semantic contents of sentences are sets of truth-supporting circumstances. Another reason is that the supplementary assumptions of the original argument are themselves highly justified.

This can be seen by looking at what many regard as the most questionable consequence of those assumptions, namely (23):[21]

(23) If i satisfies ⌜x v's that S⌝ relative to a context C (and assignment f), and if t and t′ are names, indexicals, or free variables having the same referent relative to C (and f), then i satisfies ⌜x v's that S′⌝ relative to C (and f), where S′ arises from S by substituting one or more occurrences of t′ for occurrences of t.

Many seem to think that counterexamples to this principle are easy to come by. In the case of belief ascriptions, they tend to be examples in which a competent speaker assents to S and ⌜I believe that S⌝, while dissenting from S′ and ⌜I believe that S′⌝, even though the latter arise from the former by substitution of names or indexicals with the same referent. Such cases tell against (23) only if assent and dissent are reliable guides to what is, and what is not, believed. However, dissent is not reliable in this way.[22]

A recent example of Mark Richard's makes this point quite nicely:[23]

> Consider *A*—a man stipulated to be intelligent, rational, a competent speaker of English, etc.—who both sees a woman, across the street, in a phone booth, and is speaking to a woman through the phone. He does not realize that the woman to whom he is speaking—*B*, to give her a name—is the woman he sees. He perceives her to be in some danger—a runaway steamroller, say, is bearing down upon her phone both. *A* waves at the woman; he says nothing into the phone. . . . If *A* stopped and quizzed himself concerning what he believes, he might well sincerely utter:
>
> (3) I believe that she is in danger.
> but not

(4) I believe that you are in danger.

Many people, I think, suppose that . . . [these sentences] clearly diverge in truth value, (3) being true and (4) being false. . . . But [this] view . . . is, I believe, demonstrably false. In order to simplify the statement of the argument which shows that the truth of (4) follows from the truth of (3), allow me to assume that *A* is the unique man watching *B*. Then we may argue as follows:

Suppose that (3) is true, relative to *A*'s context. Then *B* can truly say that the man watching her—*A*, of course—believes that she is in danger. Thus, if *B* were to utter

(5) The man watching me believes that I am in danger

(even through the telephone) she would speak truly. But if *B*'s utterance of (5) through the telephone, heard by *A*, would be true, then *A* would speak truly, were he to utter, through the phone

(6) the man watching you believes that you are in danger.

Thus, (6) is true, taken relative to *A*'s context.

But, of course,

(7) I am the man watching you

is true, relative to *A*'s context. [Which is not, of course to say that *A* would accept it. My addition.] But (4) is deducible from (6) and (7). Hence, (4) is true, relative to *A*'s context.

In this example, Richard is concerned with substitution of coreferential indexicals. However, the argument seems to generalize. Suppose, for example, that ⌜A believes that Ruth Barcan is F⌝ is true relative to a context. ⌜A believes that I am F⌝ should then be true relative to a corresponding context in which Ruth Barcan (i.e. Ruth Marcus) is the agent (where F is free of first person pronouns). Suppose, in fact, that Ruth utters the sentence in a conversation with someone who knows her as "Ruth Marcus". It would seem that this person can truly report ⌜A believes that she (pointing at Ruth) is F⌝, or even ⌜A believes that Ruth Marcus is F⌝. Thus, substitution of one coreferential name or indexical for another preserves truth value. Since there seems to be nothing special about this example, we have a general argument for (23).[24]

Why, then, does substitution so often provoke resistance? The answer, I think, has to do, at least in part, with the conversational purposes served by propositional attitude ascriptions. For example, suppose that Mary's neighbour, Samuel Clemens, is in the habit of soliciting her opinion of his manuscripts before sending them off to the publisher. Mary thinks they are wonderful, and regards Mr Clemens (whom she knows only under that name) as a great writer. The question is, does she think that Mark Twain is a great writer?

First consider a conversation the purpose of which is to determine Mary's opinion of various authors. The conversational participants, who use the name 'Mark Twain' to refer to the author, want to know Mary's opinion of him. I, knowing Mary's situation, report "Mary thinks that Mark Twain is a great writer." My remark seems perfectly acceptable.

However, now consider a different conversation. Mary, who is a student, has just taken a written examination, and her teacher is explaining why she failed to get a perfect score. The teacher says, "Mary did a good job, but she didn't know that Mark Twain is a writer." In the context of this conversation, the teacher's remark also seems acceptable.

But how can it be? Surely it is not the case that Mary thinks that Mark Twain is a great writer, while not knowing that Mark Twain is a writer at all.[25] To clarify this we need to distinguish between the proposition semantically expressed by a sentence relative to a context, and the information conveyed by an utterance of the sentence in a conversation. In the second conversation, the proposition semantically expressed by the propositional attitude ascription is false, even though the primary information conveyed by the utterance is true—namely, that Mary did not know that 'Mark Twain is a writer' is true; and hence was not able to answer exam questions of the sort, "What is Mark Twain's profession?" The teacher's utterance seems acceptable because the main information it conveys is correct.

This example brings out an important point about the relationship between propositional and sentential attitudes. Attitudes like asserting and believing are relations between individuals and propositions. However, often these attitudes arise in connection with attitudes towards sentences—e.g. uttering and accepting. Although propositional attitude ascriptions report relations to particular propositions, they often suggest corresponding relations to certain sentences. For example, a competent speaker of English typically (though not always) knows that 'Mark Twain is a writer' is true iff he knows that Mark Twain is a writer. Thus, it is natural that the teacher's remark should carry the meta-linguistic suggestion.

It is also natural that in many cases these suggestions should be important to the conversation. As John Perry has emphasized, sentential attitudes are often more significant for explaining action than propositional attitudes are.[26] Think again of Richard's telephone example. Suppose that a third party asks the question "Why doesn't *A* tell *B* that she is in danger?" (We assume that *A* knows his conversational partner under the name '*B*' and accepts 'You are *B*' in the context.) It is tempting to try to explain *A*'s behaviour by saying "*A* doesn't know that *B* is in danger." But this, as we have seen, is false. A better explanation is that *A* does not accept the sentence 'B is in danger'. The reason we are tempted by the propositional attitude ascription is that normally we would expect A to accept the sentence iff he thought that B was in danger. However, in this case the usual correlation between sentential and propositional attitudes breaks down. As a result, the explanation suggested by the propositional attitude ascription is correct, even though the ascription itself is false.

The general thesis, then, is that the substitutivity principle (23) is correct; and that resistance to it is based on a failure to properly distinguish the semantic information expressed by a sentence relative to a context from the information conveyed by an utterance of it in a given conversation.[27] If this is correct, then the main objection to assumptions (A2), (A4), and compositionality is eliminated, and the case against (A1*a*) is strengthened.

VI

What becomes of the difficulties in III once this assumption is given up? Taking the argument in (8) as a representative example, we see that the move from (8*b*) to (8*c*), and ultimately to (8*d*), is no longer warranted. In order to defend this as the proper response to the difficulty, I must explain how one might believe (or assert) an instance of an existential generalization, without believing (or asserting) the generalization itself. Let us focus in particular on the notion of belief. Then, what must be explained is how an individual might satisfy an open sentence ⌜x believes that R (t,t)⌝, for directly referential t, without satisfying ⌜x believes that R(t,t) and for some y, R(y,y)⌝, or ⌜x believes that for some y, R(y,y)⌝.

It should be noted that the answer is not that the agent may never have actually drawn the relevant conclusion. For the problematic derivation in (8) would proceed from true premisses to a false conclusion even if the agents were perfect logicians. Thus, there must be some deeper explanation of how a person might fail to believe the existential generalization of something he already believes.

There are two different aspects of such an explanation. The first is a metaphysical characterization of the nature of belief, specifying the facts in virtue of which belief ascriptions are true. The second is a specification of the objects of belief needed in a semantic theory. I will say a word about each.

Regarding the former, we may think of beliefs as arising from certain kinds of mental states, together with their causal relations to objects in the environment.[28] On this picture, a belief report ⌜x believes that S⌝ characterizes the agent as being in a mental state whose information content is identical with the semantic content of S in the context of the report. For example, an agent who is in a mental state appropriate for believing that a particular object is F will be correctly reported to believe that Phosphorus is F just in case the relevant part of his belief state is causally anchored to Phosphorus. Since Phosphorus is Hesperus, the agent will thereby believe that Hesperus is F.

Suppose the agent believes that Hesperus bears R to Phosphorus. On this picture, he thereby believes of a certain object o that o bears R to o. However, it does not follow that he believes the proposition that something bears R to itself. Since none of the agent's mental states has this as its information content, he does not believe it.

If we restrict our attention to cases in which the agent is a competent speaker of a language, we can make this account less abstract by letting dispositions to assent to sentences play the role of mental states. We then assume something like (24):[29]

(24) If i is a sincere, reflective, and competent speaker, then i satisfies ⌜x believes that S⌝ relative to a context C (and assignment f) iff i is disposed to assent to some sentence S′ whose semantic content in the context of assent = the semantic content of S relative to C (and f).

Let us suppose that the agent accepts ⌜R(Hesperus, Phosphorus)⌝ while rejecting ⌜R(Hesperus, Hesperus)⌝ and ⌜For some x R(x,x)⌝. An impeccable logician, the agent would accept the latter if he accepted any of its instances, ⌜R(a,a)⌝. However, he rejects all of these. Since the semantic content of one of the sentences he accepts is identical with the semantic content of ⌜R(Hesperus, Hesperus)⌝, he believes that Hesperus bears R to Hesperus even though he would not express his belief this way. Since the semantic content of ⌜For some x R(x,x)⌝ is not identical with the content of any sentence he is disposed to accept, he does not believe that something bears R to itself. Thus, there is a principled way of blocking the move from (8*b*) to (8*d*).

What we need now is a conception of semantic content capable of incorporating this point. Given that the move from (8*c*) to (8*d*) is unproblematic, we need a conception that blocks the move from (8*b*) to (8*c*) by assigning different semantic contents to the complement sentences in these examples. This requires the introduction of structure into contents.

First consider simple sentences:

(25*a*) R(Hesperus, Phosphorus)
(25*b*) R(Hesperus, Hesperus)
(25*c*) R(Hesperus, itself).

Regimenting a bit, we can think of the semantic contents of these examples as being identical with that of certain canonical representations:

(26*a*) R(h,p)
(26*b*) R(h,h)
(26*c*) [λx R(x,x)] h.

Where o is the referent of 'Hesperus' and 'Phosphorus' the content of (*a*) and (*b*) is, in effect, ⟨⟨o,o⟩, the two place property R⟩; the content of (*c*) is ⟨⟨o⟩, the one place property of bearing R to oneself⟩.[30]

Accepting (*a*) leads to a belief whose object is the first of these semantic contents; accepting (*c*) leads to a belief whose object is the second such content. Accepting (*b*) typically leads to a belief in both. The reason for this has to do with the transparent linguistic relationship between (*b*) and (*c*). A competent speaker who accepts one will normally be disposed to accept the other, thereby acquiring both beliefs.[31]

Thus, a speaker who satisfies (27*a*) will standardly satisfy both (27*b*) and (27*c*):

(27*a*) x accepts 'R(Hesperus, Hesperus)'
(27*b*) x believes that R(Hesperus, Hesperus)
(27*c*) x believes that R(Hesperus, itself).

However, not everyone who satisfies (27*b*) satisfies (27*c*). Whether or not the latter is satisfied will depend on the manner in which the agent believes that Hesperus bears R to Hesperus. If he believes it in virtue of accepting a sentence of the form 'R(a,a)' then he can be expected to believe that Hesperus bears R to itself. However, if he believes it in virtue of accepting a sentence of the form 'R(a,b)', then (27*c*) may not be satisfied.

The same point holds for (27*d*):

(27*d*) x believes that for some y, R(y,y).

A sincere, reflective, competent speaker who accepts ⌜R(a,a)⌝ will typically be disposed to accept ⌜For some y R(y,y)⌝, and thereby believe that which it expresses. However, someone who accepts ⌜R(Hesperus, Phosphorus)⌝ may satisfy (27*b*) without satisfying (27*d*).

In order to reflect this in a semantic theory we must extend our account of structured semantic contents from atomic sentences to compound sentences of arbitrary complexity. This raises the question of how much structure is needed. Where S is an atomic sentence consisting of an n-place predicate plus n occurrences of directly referential terms, its structured semantic content consists of the content of the predicate plus the content of each term occurrence. There are two ways of thinking of this—as a complex made up of the semantic contents of all occurrences of its semantically significant parts, or as a complex made up of the contents of all occurrences of its directly referential terms, plus the content of whatever else is left over. In the case of atomic sentences, these characterizations come to the same thing. However, they generalize in different ways. The first leads to a conception of the semantic contents of sentences as structured Russellian propositions, the second to a conception of contents as partially structured intensions.

For simplicity let us consider the semantic contents of sentences in a first order language with lambda abstraction, a belief predicate, and a stock of semantically simple singular terms, all of which are directly referential. On the Russellian account, the semantic content of a (free) variable v relative to an assignment f of individuals to variables is f(v), and the semantic content of a closed (directly referential) term, relative to a context, is its referent relative to the context. The semantic contents of n-place predicates are n-place properties and relations. The contents of

'&' and '—' are functions, CONJ and NEG, from truth values to truth values.[32]

Variable binding operations, like lambda abstraction and existential quantification, can be treated in a number of ways. One of the simplest, semantically, involves the use of propositional functions in place of complex properties as propositional constituents corresponding to certain compound expressions.[33] On this approach, the semantic content of ⌜[λx Rx,x]⌝ is the function g from individuals o to propositions that attribute the property expressed by R to the pair ⟨o,o⟩. ⌜∃x Rx,x⌝ can then be thought of as "saying" that g assigns a true proposition to at least one object.

(28) uses these ideas to assign Russellian propositions to sentences:

(28*a*) The proposition expressed by an atomic formula ⌜$Pt_1, \ldots, t_n$⌝ relative to a context C and assignment f is ⟨⟨$o_1, \ldots, o_n$⟩, P*⟩, where P* is the property expressed by P, and o_i is the content of t_i relative to C and f.

(28*b*) The proposition expressed by a formula ⌜[λvS] t⌝ relative to C and f is ⟨⟨o⟩,g⟩, where o is the content of t relative to C and f, and g is the function from individuals o′ to propositions expressed by S relative to C and an assignment f′ that differs from f at most in assigning o′ as the value of v.

(28*c*) The propositions expressed by ⌜–S⌝ and ⌜S&R⌝ relative to C and f are ⟨NEG, Prop S⟩ and ⟨CONJ, ⟨Prop S, Prop R⟩ ⟩ respectively, where Prop S and Prop R are the propositions expressed by S and R relative to C and f, and NEG and CONJ are the truth functions for negation and conjunction.

(28*d*) The proposition expressed by ⌜∃v S⌝ relative to C and f is ⟨SOME, g⟩, where SOME is the property of being a non-empty set, and g is as in (*b*).

(28*e*) The proposition expressed by ⌜t believes that S⌝ relative to C and f is ⟨⟨o, Prop S⟩, B⟩, where B is the belief relation, o is the content of t relative to C and f, and Prop S is the proposition expressed by S relative to C and f.

(28*f*) The proposition expressed by a sentence (with no free variables) relative to a context C is the proposition it expresses relative to C and every assignment f.

In stating clause (*d*), I have departed slightly from Russellian ideas in favour of a suggestion by Nathan Salmon. A purely Russellian approach would treat SOME as the property of being a propositional function that is "sometimes true". However, since the existential quantifier is an extensional operator, it seems more natural that it should express a property of the extension of its operand (rather than a property of the propositional constituent expressed by the operand, as in the case of 'believe'). On this formulation, ⟨SOME, g⟩ is true relative to a circumstance E iff the set of objects in E that g maps onto propositions true in E is non-empty.

This discussion of truth conditions brings up an important point. Propositional contents do not replace truth-supporting circumstances in a semantic theory; rather, they supplement them with a new kind of semantic value. On this view, the meaning of an expression is a function from contexts to propositional constituents. The meaning of a sentence is a compositional function from contexts to structured propositions. Intensions (and extensions) of expressions relative to contexts (and circumstances) derive from intensions (and extensions) of propositions and propositional constituents. These, in turn, can be gotten from a recursive characterization of truth with respect to a circumstance, for propositions.

For this purpose, we let the intension of an n-place property be a function from circumstances to sets of n-tuples of individuals (that instantiate the property in the circumstance); we let the intension of an individual be a constant function from circumstances to that individual; and we let the intension of a one place propositional function g be a function from circumstances E to sets of individuals in E that g assigns propositions true in E. Extension is related to intension in the normal way, with the extension of a proposition relative to a circumstance being its truth value in the circumstance, and its intension being the set of circumstances in which it is true (or, equivalently, the characteristic function of that set). Truth relative to a circumstance is defined as follows:

(29*a*) A proposition $\langle\langle o_1, \ldots, o_n\rangle, P^*\rangle$ is true relative to a circumstance E iff the extension of P* in E contains $\langle o_1, \ldots, o_n\rangle$.

(29*b*) A proposition $\langle\langle o\rangle, g\rangle$ is true relative to E (where g is a one place propositional function) iff o is a member of the extension of g in E (i.e. iff g(o) is true in E).

(29*c*) A proposition ⟨NEG, Prop S⟩ is true relative to E iff the value of NEG at the extension of Prop S in E is truth (i.e. iff Prop S is not true in E). A proposition ⟨CONJ, ⟨Prop S, Prop R⟩⟩ is true relative to E iff the value of CONJ at the pair consisting of the extension of Prop S in E and the extension of Prop R in E is truth (i.e. iff Prop S and Prop R are true in E).

(29*d*) A proposition ⟨SOME, g⟩ is true relative to E (where g is as in (*b*)) iff the extension of g in e is non-empty (i.e. iff g(o) is true relative to E for some o in E).

(39*e*) A proposition ⟨⟨o, Prop S⟩, B⟩ is true relative to E iff ⟨o, Prop S⟩ is a member of the extension of B in E (i.e. iff o believes Prop S in E).

According to this theory, the propositions expressed by the complements of (8*b*) and (8*c*) are (8*b**) and (8*c**):

(8*b**) ⟨CONJ, ⟨⟨⟨ 'Hesperus', Hesperus⟩, the reference relation⟩, ⟨⟨'Phosphorus', Hesperus⟩, the reference relation⟩⟩⟩

(8*c**) ⟨CONJ, ⟨(8*b**), ⟨SOME, g⟩⟩⟩ (Where g is the function which assigns to any object o the proposition about o corresponding to the proposition 8*b** about Hesperus.)

Although the circumstances supporting the truth of these propositions are the same, the propositions themselves are different. Thus we no longer have the result that anyone who believes the proposition expressed by the complement of (8*b*) thereby believes the proposition expressed by the complement of (8*c*). The argument in (8) is, therefore, blocked and the problematic conclusion avoided. Similar results hold for the other arguments in III.

However, this is not the only way these results can be achieved. One of the striking features of Russellian propositions is that they encode a good deal of the syntactic structure of the sentences that express them. Sentences that are negations, conjunctions, or quantifications express propositions which are themselves negative, conjunctive, or quantificational in structure.[34] Although this systematic assignment of structure to semantic contents is appealing, it goes beyond what is required by the interaction of propositional attitudes and directly referential singular terms exhibited in III.

In each of the problematic arguments, the agent accepts, or assertively utters, a sentence of the form (30*a*), but fails to accept, or assertively utter, a corresponding sentence of the form (30*b*) (which is true in the same circumstances as (30*a*)):

(30*a*) S(t, t′)

(30*b*) S(t, t′) & R.

In each case, the agent would accept, or assertively utter, (30*b*) if he knew that the directly referential terms t and t′ had the same content (and he continued to accept (30*a*)). However, he does not know that they have the same content. In order to focus on the special difficulties created by this sort of ignorance, let us suppose, for the sake of argument, that the agent is otherwise semantically omniscient. Thus he knows, for any two expressions not containing directly referential terms, whether or not they have the same intension.

In particular, he knows this about (31*a*) and (31*b*):

(31*a*) λv,v′[S(v,v′)]

(31*b*) λv,v′[S(v,v′) & R].

If he thought that these expressions had the same intension, then his attitude towards (30*a*) and (30*b*) would be the same—he would either accept them both or reject them both. Since, in fact, he accepts one and rejects the other, it follows that (31*a*) and (31*b*) have different intensions.

This means that whenever an argument of the sort presented in III can be constructed, its problematic conclusion can be blocked by taking the semantic content of a sentence to be a complex consisting of intensions of all occurrences of its directly referential singular terms, plus an intension determined by the remainder of the sentence. The idea can be carried out using a standard style definition of truth with respect to a context and circumstance. Such a definition allows one to associate both a standard intension and a partially structured intension with every object language sentence. Standard intensions of sentences can be taken to be sets of truth-supporting circumstances. Partially structured intensions are complexes made up in part of the intensions of directly referential terms. If a sentence contains no such terms, then its partially structured intension is identified with its standard intension.

We can make this more precise as follows: let us call an occurrence of a singular term in a sentence S a *structurally sensitive occurrence* iff it is a free occurrence of a variable in S or it is an occurrence of a (constant) directly referential term.[35] Let $\ulcorner\lambda v_1, \ldots, v_n\ S'\urcorner$ arise from S by prefixing $\ulcorner\lambda v_1, \ldots, v_n\urcorner$ and replacing each structure sensitive occurrence of a singular term in S with a variable new to S, distinct variables for distinct occurrences, v_i replacing the ith such occurrence. The extension of $\ulcorner\lambda v_1, \ldots, v_n\ S'\urcorner$ relative to an assignment f, context C, and circumstance E is taken to be the function from n-tuples $\langle o_1, \ldots, o_n\rangle$ to truth values of S′ relative to f′, C, and E, where f′ is just like f except (at most) for assigning o_i, as the value of v_i, for each i. Standard intension is determined from extensions in the normal way. For any (open or closed) sentence S, the *partially structured intension* of S relative to an assignment f and context C is $\langle\langle[t_1], \ldots, [t_n]\rangle, [\lambda v_1, \ldots, v_n S']\rangle$, where $[t_i]$ is the intension of the ith structure sensitive occurrence of a singular term in S, relative to f and C, and $[\lambda v_1, \ldots, v_n S']$ is the intension of $\ulcorner\lambda v_i, \ldots, v_n S'\urcorner$, relative to f and C. (Closed sentences have the same partially structured intensions—with respect to a context—relative to all assignments.) An individual i satisfies an open sentence ⌜x believes that S⌝, relative to f, C, and E, iff in E i bears the belief relation to the partially structured intension expressed by S relative to f and C.[36]

Conceptually, this approach lies somewhere between the Russellian theory and the familiar truth-supporting circumstance conception. Like the Russellian theory, it takes propositions to be structured complexes which are both the semantic contents of sentences and the objects of propositional attitudes.[37] However, unlike the Russellian theory, the constituents of these "propositions" are intensions extractable from a conventional truth definition. Moreover, the resulting "propositions" are only partially structured.

For example, the partially structured contents of the complements of (8*b*) and (8*c*) are:

(8*b*#) ⟨⟨'Hesperus', Hesperus, 'Phosphorus', Hesperus⟩, R′⟩ (where R′ is the intension corresponding to the 4-place relation of x's referring to y and z's referring to v).[38]

(8*c*#) ⟨⟨'Hesperus', Hesperus, 'Phosphorus', Hesperus⟩, R″⟩(where R″ is the intension corresponding to the 4-place relation of x's referring to y and z's referring to v and there being a common referent of x and z).

Since R′ is not identical with R″, these contents are different. The move from (8*b*) to (8*c*) is, therefore, blocked and the problematic conclusion (8*d*) is avoided. Corresponding results hold for other arguments of this type, including those in III.

This approach represents a theoretically minimum response to the difficulties in III. As such it allows us to establish a minimum positive result about the relationship between direct reference and propositional attitudes, corresponding to the impossibility result, (13):[39]

(32) If direct reference is legitimate and (some) propositional attitude verbs have a relational semantics ((A4) plus (A2)), then (assuming compositionality and distribution over conjunction) the semantic content of a sentence, relative to a context and assignment of values to variables, must encode at least as much structure as is determined by occurrences of its directly referential singular terms (including free variables).

Both structured Russellian propositions and partially structured intensions satisfy this requirement.

VII

How then might we decide between these two conceptions of semantic content? Considerations involving the interaction of propositional attitudes and directly referential singular terms will, I believe, take us no further. However, other considerations will.

The first of these involves related expressions which allow the construction of arguments corresponding to those in III. For example, if K and K′ are natural kind terms with the same semantic content, the potentially false (12*b*′) can be derived from the potentially true (12*a*′) by an argument paralleling the original (12):

(12*a*′) x believes (asserts) that the G is a K and if the G is a K′, then S (where S is any sentence and ⌜the G⌝ is any description).

(12*b*′) x believes (asserts) that S.

Both this argument and the original (12) are blocked by requiring the semantic content of a sentence to encode at least as much structure as is determined by occurrences of its directly referential singular terms, *plus* its natural kind terms.

This conclusion can be extended to include every kind of

expression that is relevantly similar to directly referential singular terms and natural kind terms. The relevant feature, I suggest, is one that involves linguistic competence—in the sense that linguistic competence is important for determining what is said or believed by a speaker from what is assertively uttered or accepted by the speaker. If it is possible for a competent speaker to fail to recognize cases in which expressions of type T have the same semantic content, then it will be possible to use these expressions to construct arguments of the kind given in III. Blocking these arguments requires ensuring that the structure encoded in semantic contents includes that determined by occurrences of expressions of type T.

This line of reasoning leads to the encoding of more and more structure into semantic contents. However, it might be thought that at least some expressions—including logical constructions plus certain predicates—remain immune from such considerations.[40] If S is a sentence containing only such expressions, then its semantic content, on the partially structured intension approach, will just be a standard intension. If S contains only such expressions plus directly referential singular terms, then its semantic content, on this approach, will be a partially structured intension in the original sense. But this is still problematic.

The difficulties posed by propositional attitude ascriptions for truth conditional approaches to semantics are not limited to cases arising from directly referential singular terms and their ilk. For example, if truth-supporting circumstances are metaphysically possible worlds, then the partially structured intension approach will assign the same semantic contents to the (*a*) and (*b*) sentences in the following examples:

(33*a*) First order logic is complete.

(33*b*) First order logic is undecidable.

(34*a*) First order logic is decidable.

(34*b*) First order logic is decidable and S (for unrelated S).

However, in both cases, many have believed or asserted that which is expressed by (*a*) without believing or asserting that which is expressed by (*b*).

Switching to a conception in which truth-supporting circumstances are logically possible worlds only shifts attention to a more restricted, but similarly problematic, class of cases. Like Frege of the *Grundgesetze*, many of us have had the misfortune of satisfying ⌜x asserts (believes) that I⌝, for a some logically impossible I, without thereby satisfying ⌜x asserts (believes) that I&S⌝, or ⌜x asserts (believes) that S⌝, for unrelated S.

The problem is, I believe, inherent, in the truth conditional approach, and, hence, cannot be solved by weakening constraints on truth-supporting circumstances still further. For example, consider a system like that of *Situations and Attitudes*, in which truth-supporting circumstances may be metaphysically impossible, incomplete, and inconsistent in the sense defined in II. In such a system, logically equivalent sentences are often assigned different semantic contents, which may be the objects of different propositional attitudes. As with all such approaches, however, the system incorporates principles like (7*a*–*e*), which can be extracted from standard, recursive treatments of logical constructions. Inevitably, sentences involving multiple constructions of this kind require psychologically non-trivial computations to determine their "semantic contents". Thus, one can always find psychologically non-equivalent sentences which are true in the same circumstances, and, hence, are assigned the same content.

One simple example of this kind is given in (35):

(35*a*) C[the x: Ax] & D[the x: Cx] & C[the x: Bx]

(35*b*) A[the x: Bx v Cx] & B[the x: Dx & Cx] & D[the x: Ax v Cx]

(35*c*) B[the x: Ax] & C[the x: Bx] & D[the x: Cx].

Although these sentences are assigned the same semantic content by corollaries (7*a*, *b*, *e*) of (A1*a*), it takes a modest amount of calculation to determine this. Not all agents of propositional attitudes are adept at such calculations. Thus, it is possible to find agents who are willing to accept, or assertively utter, one of these sentences at a certain time, but not the others. Such agents believe, or assert, that which is expressed by the sentence they accept, or assertively utter. However, it is counterintuitive to suppose that they must thereby believe, or assert, what the other sentences express. The Russellian conception of propositions allows one to respect this intuition; the truth-supporting circumstance approach does not.

A related point involves the relationship between propositional attitudes and conjunction. Surely, anyone who believes that (35*a*), or believes that (35*c*), believes that (36):[41]

(36) C[the x: Bx].

However, this does not seem to be so with (35*b*). The reason for this difference is that in one case the move is from a belief in a conjunction to a belief in a conjunct, whereas in the other it is not. Although many logical operations do not preserve belief, it would seem that simplification of conjunction does.

In fact, I think his observation about conjunction and belief is correct. However, it has far-reaching theoretical significance that belies its widespread acceptance. Let us suppose that what are believed are semantic contents of sentences. On the truth-supporting circumstance approach, these contents never have conjunctive structure. At best, they are partially structured intensions, which reflect the structure determined by occurrences of directly referential terms (and related expressions), but obliterate other logical structure. Thus, on this approach, there is no more reason to think that anyone who believes that (35*a*) (or (35*c*)) believes that (36) than there is to think that anyone who believes that (35*b*) does.

Proponents of the truth-supporting circumstance approach can, of course, countenance the move from belief in that which is expressed by a conjunction to belief in that which is expressed by the conjuncts. Indeed, they standardly do. However, the price to be paid is that of countenancing the move from ⌜x believes that S⌝ to ⌜x believes that S′⌝ whenever the set of circumstances suporting the truth of S is included in the set of circumstances supporting the truth of S′. But this just substitutes the generation of unwanted inferences for the failure to capture one that is desired. In short, the truth-supporting circumstance approach does not provide the right options.[42]

The Russellian approach offers a welcome contrast. Given the intuition that whenever an individual satisfies ⌜x believes that A & B⌝ he also satisfies ⌜x believes that A⌝ and ⌜x believes that B⌝, the Russellian approach supplies a plausible explanation. Since objects of belief reflect the logical structure of sentences used to report those beliefs, whenever a belief is correctly reported using a conjunction the agent will believe a conjunctive proposition which includes the propositions expressed by the conjuncts as constituents. Since these constituent propositions are, so to speak, before his mind, no computation is required in order for him to arrive at beliefs in the conjuncts.

We can think of this somewhat less metaphorically as follows: to believe a conjunctive proposition ⟨CONJ, ⟨P, Q⟩⟩ is to be in a belief state whose constituents correspond to its three main components. In the case of CONJ, this correspondence is, presumably, functional. A belief state constituent C represents CONJ only if an individual who is in a "conjunctive belief state" S, in which C relates constituent belief states S1 and S2, is also in—or disposed to be in—S1, and S2. Thus, anyone who believes a conjunction believes both conjuncts.

The point to notice is that with propositions as semantic contents this result does not generalize in unwanted ways. Even though structured propositions determine truth-supporting circumstances, there is no reason to suppose that just because an agent bears the belief relation B to a proposition P, he must also bear B to Q whenever the class of truth-supporting circumstances for P is identical with, or a subclass of, the class of truth-supporting circumstances for Q.

There are, then, good reasons not only for rejecting a strict truth-supporting circumstance conception of semantics, but also for adopting a Russellian approach. The reasons I have stressed rest on commonplace intuitions and assumptions about propositional attitudes. There are, of course, those who regard the attitudes as ill-behaved and problematic, and would, therefore, not accept such intuitions and assumptions. In my opinion, such pessimism is unwarranted.

If I am right, a major reason why propositional attitudes have often seemed intractable is that the basic features of strict truth-theoretic semantics have been incompatible with elementary facts about them. The introduction of structured Russellian propositions, which determine, but are not determined by, sets of truth-supporting circumstances, has the potential to change that.

Notes

* This article grew out of work originating in my critique, 'Lost Innocence', *Linguistics and Philosophy* 8. 1 (1985), 59–71, of J. Barwise and J. Perry, *Situations and Attitudes* (MIT Press, 1983). It was written in 1983–4 while on leave from Princeton University on the Class of 1936 Bicentennial Preceptorship, and while a guest, first, of The Syntax Research Center at the University of California, Santa Cruz, and, later, of the philosophy department of the University of Washington. Portions of it provided the basis for talks at the University of California at Berkeley, Riverside, and Santa Cruz (1983–4); the

University of Illinois (1985); North Carolina State University (1985); the University of Pennsylvania (1985); Princeton University (1984); Simon Fraser University (1983); Stanford University (1984); and the Pacific Division Meetings of the American Philosophical Association (1985). A shortened version of the paper, adapted from the APA talk, will appear in *Themes from Kaplan*, edited by Joseph Almog, John Perry, and Howard Wettstein (Oxford University Press, 1988).

I have benefited considerably in the development of several important points from extensive discussion and correspondence with Joseph Almog, David Kaplan, and Nathan Salmon. I have also profited from discussion with Ali Akhtar Kazmi, Julius Moravcsik, and Mark Richard.

1. I assume here, and in what follows, that the semantic content of the complement sentence in a propositional attitude ascription is compositionally determined from the semantic contents of its parts.
2. For example, the properties of being human and of not being human are complements of one another. I will assume that every property has a (unique) complement and that P is the complement of Q iff Q is the complement of P.
3. The idea of thinking of abstract situations as resulting from relaxing constraints on Carnapian state descriptions was suggested to me by David Kaplan.
4. Thus, the content of an existential generalization is a superset of the contents of instances from which it follows. It should be noted that no formal treatment of existential quantification is provided in *Situations and Attitudes*. Nevertheless, (7*c*) accords well with the leading ideas of that work. (7*a*), (7*b*), (7*d*) and (7*e*) are explicitly endorsed.
5. The compositional principle I will appeal to may be understood as applying to sentences free of quotation and opacity-producing operators:

   If S and S′ are non-intensional sentences with the same grammatical structure, which differ only in the substitution of constituents with the same semantic contents (w.r.t. their respective contexts and assignments of values to variables), then the semantic contents of S and S′ will be the same (w.r.t. those contexts and assignments).

   This principle is presupposed in standard versions of truth-conditional semantics, and is itself a corollary of assumption (A1*a*).
6. So long as they validate (7*a*) and (7*e*). This continues to hold when any two place relation replaces identity.
7. One might, of course, try to avoid this result by tampering with the semantics of definite descriptions. For example, one might try substituting the unlovely (7*e*′) for (7*e*).

   (7*e*′) The semantic content of ⌜F[the x: Gx]⌝ (relative to a context) is the set of circumstances E such that there is at least one object o in E which is both an F-er and a G-er in E; and moreover, for any other object o′, if o′ is a G-er in E, then o = o′ and o′ = o in E, and, more generally, o and o′ have exactly the same properties (and stand in the same relations to the same objects) in E.

   One drawback of this from the point of view of a system like that of *Situations and Attitudes* is that it gives up the view that definite descriptions determine partial functions from circumstances to objects that uniquely satisfy the descriptions in those circumstances. Since this feature of definite descriptions is used extensively in *Situations and Attitudes*, it is not clear that Barwise and Perry would be willing to replace (7*e*) with (7*e*′). In any case, such a move would do nothing to remove the problem posed by (9*a*′).

8. The derivation in (12) depends on the assumption that if E supports the truth of P and also supports the truth of ⌜P → Q⌝, then E supports the truth of Q. This will hold if truth-supporting circumstances are logically possible words and E supports the truth of a material conditional whenever it supports the truth of the consequent or fails to support the truth of the antecedent. When circumstances are allowed to be partial and inconsistent, the situation is no longer straightforward. For example, the system in *Situations and Attitudes* provides no treatment of conditionals, and so is not subject to the argument based on (12).
9. Here, (A1)–(A4) must be understood as relativizing semantic content and associated truth conditions to both contexts and assignments of values to variables:

   (A1*a*) The semantic content of a sentence (relative to a context C and assignment f) is the collection of circumstances supporting its truth (as used in C with respect to f).

   (A2) An individual i satisfies ⌜x v's that S⌝ (relative to C and f) iff i bears R to the semantic content of S (relative to C and f).

   (A3) For many propositional attitude verbs (including 'say', 'assert', and 'believe') if i satisfies ⌜x v's that P & Q⌝ (relative to C and f), then i satisfies ⌜x v's that P⌝ and ⌜x v's that Q⌝ (relative to C and f).
10. S. A. Kripke, 'A Puzzle about Belief' in A. Margalit (ed.), *Meaning and Use* (Reidel: Dordrecht, 1979), 239–75; also in this volume.
11. See 'A Puzzle about Belief' p. 268 (133–4 of this volume) and n. 42 for relevant discussion.
12. (A2′) is a consequence of (A2), but not vice versa. If (A2) is true, then R* in (A2′) can be taken to be a relation that an individual i bears to a content-character pair, ⟨y,z⟩, iff i bears the relation R of (A2) to the content y. However, (A2′) might be true even if substitution of complement sentences with the same content failed to preserve truth value, in which case (A2) would be false.
13. The basis for these reductios is partially prefigured in 'A Puzzle about Belief', pp. 257–8, 262 (122, 126–7 of this volume).
14. In the case of (10), one can use (7*e*) to derive ⌜x believes that S&P⌝ from ⌜x believes that S⌝, where S is (10*c*) and P is *any sentence at all*. See, however, the qualification in n. 7.
15. See M. Richard, 'Grammar, Quotation, and Opacity', *Linguistics and Philosophy*. 9.3 (1986), 383–403.
16. See 'A Puzzle about Belief', n. 43 for relevant discussion.
17. I am indebted to Joseph Almog for suggesting that I reconstruct my argument using this sort of weakening of (A2).
18. 'A Puzzle about Belief', pp. 260–3 (125–7 of this volume). If the properties are not required to pick out a unique referent (e.g. if the property of being a famous Roman is the one associated with both 'Cicero' and 'Tully'), then problematic substitution will go through even when the property is "purely qualitative". See 'A Puzzle about Belief', n. 9.

19. There is another respect in which (A2) and its weakened counterparts may give a misleading impression of strength. They may suggest that the arguments against (A1*a*) rely crucially on assumptions about the semantics of sentences of the form ⌜x v's that S⌝. In fact, such sentences are dispensable.

    There are two leading ideas behind the various versions of (A2). The first is that propositional attitudes like saying, asserting, and believing are relations to things that are said, asserted, and believed. The second is that these things are said, or semantically expressed, by sentences. If these ideas are correct, then the arguments against (A1*a*) can be reconstructed—either directly, in terms of *what sentences say*, or indirectly, using (1) and, if desired, (A3′) to derive conclusions about *what speakers say*.

    (A3′) If x says (or asserts) that which is said (expressed) by a conjunction in a context C, then x says (or asserts) that which is said (expressed) by each conjunct in C.

    Using these principles, one can derive the incorrect conclusion that x has said (or asserted) that which is expressed by ⌜For some y Ryy⌝ from the premiss that x has assertively uttered ⌜R(Hesperus, Phosphorus)⌝ or ⌜R(Londres, London)⌝
20. The same point could be made using other logical constructions—for example, indefinite descriptions—in place of existential quantification in the complement sentence.
21. Where S is free of quotation and related constructions.
22. The other main type of putative counterexample to (23) involves cases in which a competent speaker assents to (translations of) ⌜n is F⌝ and ⌜m is not F⌝ in a context in which n and m are coreferential names or indexicals. With (16*a*) plus translation one gets the result that the agent satisfies ⌜x believes that n is F⌝ and ⌜x believes that m is not F⌝. Substitutivity then results in the ascription of contradictory beliefs, which is sometimes thought to be objectionable in light of the fact that the agent may have made no logical mistakes.

    However, Kripke's example of puzzling Pierre shows that this is not a compelling criticism of (23). As we have seen, ascriptions of contradictory statements and beliefs can be derived from (16) plus translation, without any appeal to substitutivity. Moreover, the inconsistency is genuine. Kripke's Pierre really does say and believe both that London is pretty and that London is not pretty. But then, if the statements and beliefs of even the best reasoner can be inconsistent without his being in a position to recognize it, the mere fact that the substitutivity principle can sometimes be used to arrive at ascriptions of such inconsistency does nothing to discredit it.
23. M. Richard, 'Direct Reference and Ascriptions of Belief', *Journal of Philosophical Logic* 12 (1983), 425–52, esp. 439–41 (184–6 of this volume).
24. Richard himself does not go this far. For one thing, his semantics for belief ascriptions is silent about sentences containing proper names. More important, however, is a weakening of (23) involving complement sentences containing two or more occurrences of indexicals and/or free variables. Let t and t′ be two such terms which have the same content relative to a context C and assignment f. According to Richard, if (i) is true relative to C, f, and a circumstance E, then (ii) must be true relative to C, f, and E, but not vice versa:

    i. x believes that S (t, t)
    ii. x believes that S (t, t′).

Both this conclusion and the semantic system that leads him to it are, in my opinion, incorrect. Nevertheless, there is an important truth underlying Richard's observations. This truth (first suggested to me by Nathan Salmon) is brought out by (iii):

iii (*a*) x believes that t is not identical with t′
(*b*) x believes that t is not identical with t
(*c*) x believes that t is not identical with itself
(*d*) x believes that t is non-self-identical.

It seems evident that (*a*) can be true when (*d*) is not. The reason for this is that believing the latter involves attributing the property of non-self-identity to an object, whereas believing the former does not. In light of this, one must block either the move from (*a*) to (*b*), or the move from (*b*) to (*c*) and (*d*).

Richard selects the first of these. According to him,

iv. x believes that S

is true only if the agent believes the proposition (semantic content) expressed by S (relative to the context and assignment). Moreover, the complements of (*a*) and (*b*) express the same proposition. Nevertheless, Richard holds that (*a*) can be true when (*b*) is false. The reason for this is that on his semantics a belief ascription of the form (iv) not only reports what proposition is believed, but also places constraints on the sentence acceptance of which is responsible for the agent's belief. In the case of (*b*), the agent must hold the belief in virtue of accepting a sentence containing occurrences of directly referential terms with the same character. (If the account were extended to names it would be more natural to require two occurrences of the same term.) In the case of (*a*), this is allowed, but not required.

One problem with this account is that it is too restricted. Whatever may be the case regarding ascriptions of the form (iv), some belief ascriptions express straightforward relations to propositions:

v (*a*) x believes the proposition expressed at the bottom of page 437
(*b*) The proposition expressed at the bottom of page 437 is the proposition that P
(*c*) Therefore, x believes the proposition that P.

Given the admission that ⌜t is not identical with t′⌝ and ⌜t is not identical with t⌝ express the same proposition, one can use examples of the form (v) to reinstate the very problems that the non-relational semantics of (iv) was designed to avoid.

It seems to me that a better approach is to take all belief ascriptions (with the possible exception of belief *de se*) to express relations to propositions (semantic contents), but to distinguish the proposition expressed by the complements of (iii*a*) and (iii*b*) from the proposition expressed by the complements of (iii*c*) and (iii*d*). In this way, one can block the move from (*b*) to (*c*), while preserving (23). An account of this kind is presented in VI below.

I am grateful to Mark Richard and Nathan Salmon for discussions of the issues in this note.

25. It might be thought that a theory that took names to be disguised descriptions

(associated with them by the speaker) could render the remarks in the two conversations true by appealing to a difference in scope. But this will not work. If the description associated by the teacher with the name 'Mark Twain' is something like 'the author of *The Adventures of Tom Sawyer* and *Huckleberry Finn*', then the teacher's remark will be false no matter what the scope of the description.

26. J. Perry, 'Frege on Demonstratives', *The Philosophical Review*, 86. 4 (1977), 474–97 and 'The Problem of the Essential Indexical', *Nous* 13 (1979), 3–21 (included in this volume, ch. VI).
27. This thesis has recently been championed by several philosophers, most notably Nathan Salmon in *Frege's Puzzle* (MIT Press, 1986). Although I developed the arguments given above independently, I have profited from Salmon's work on the topic.
28. See *Situations and Attitudes* for an articulation of this view.
29. Although this principle is a useful heuristic, it should not be regarded as an analysis of belief. Its most obvious limitation is that it does not apply to believers who are not language users. Even when applied to language users it must be restricted to cases in which the agent, i, the sentence S′, and its semantic content (in the context of assent) stand in a certain (as yet not fully analysed) recognition relation. I have in mind examples like 'Newminister 1', in which a proper name is introduced by a reference-fixing description 'the first Tory Prime Minister of Britain elected in the 21st century'. (The example parallels the 'Newman 1' example discussed in K. Donnellan 'The Contingent A Priori and Rigid Designators' in P. A. French, T. E. Uehling, Jr., and H. K. Wettstein (eds.), *Contemporary Perspectives in the Philosophy of Language* (University of Minnesota Press, 1977)). In such a case, the sentence 'Newminister 1 will be conservative' may express a singular proposition involving a certain individual. However, assent to the sentence by a competent speaker is not sufficient for belief in that proposition. Intuitively, the manner in which the sentence presents the proposition to the agent is too indirect for assent to indicate belief.

It should be noted that the cases discussed in the text ('Hesperus'/'Phosphorus', 'London'/'Londres', etc.) are not like this. In these cases, the agents are acquainted with the referents, they associate names with them, and they grasp the propositions expressed by sentences containing the names. What they do not do is recognize that the same referents are associated with different names, and that the same propositions are expressed by different sentences. But that is not required in order for assent to the sentences to indicate belief in the propositions they express.

Similar points can be made about assertion (except that here the principles involving assent come closer to providing an actual analysis).

(i) An individual i satisfies ⌜x says (asserts) that S⌝ relative to a context C (assignment f) and circumstance E, if there is a sentence S′ and context C′ corresponding to E with i as agent, such that i assertively utters S′ in C′, and the content of S′ in C′ = the content of S in C (relative to f).

(ii) An individual i satisfies ⌜x says (asserts) that S⌝ relative to a context C (assignment f) and circumstance E iff there are sentences S′ and S″, and a context C′ corresponding to E with i as agent, such that i assertively utters

S′ in C′, S″ is readily inferable from S′, and the content of S″ in C′ = the content of S in C (relative to f).

If one takes (ii) to be a reasonable approximation of the notion of assertion, one can use it in place of (24) to construct arguments and explanations involving assertions corresponding to those in the text involving beliefs. (Note that (ii) allows contents not expressed by the sentence uttered to be asserted when, but only when, the conversational participants "can't miss them".)

30. The idea for this lambda-treatment of (25*c*) was suggested to me by Nathan Salmon.
31. Except in situations like Kripke's Paderewski example, in which the agent misconstrues two tokens of the same name (referring to the same individual) for tokens of different (but phonologically identical) names of different individuals ('A Puzzle about Belief', 265–6; 130–1 this volume). In such cases an agent might accept (*b*) without accepting (*c*), or believing what it expresses.
32. On this treatment, '&' and '—' are like directly referential terms in that their semantic contents = their extensions. This is not crucial to the general Russellian conception, which could just as well take the contents of these expressions to be entities—call them operations—which bear a relation to functions analogous to that borne by properties to the objects they apply to. It is even possible to take the contents of truth functional operators to be properties of truth values. The differences between these alternatives do not affect the present discussion.
33. I am indebted to David Kaplan for this Russellian suggestion.
34. It is, of course, possible for sentences of one form to express propositions of another form, as happens in some cases of stipulative definition.
35. I retain here the simplifying assumption that all directly referential terms in the object language are semantically simple.
36. This account has two precursors. The first is the introduction of structured meanings (characters) in Richard's 'Direct Reference and Ascriptions of Belief'. The second is a somewhat different use of structured intensions suggested by David Kaplan (personal correspondence) in response to Richard. The account in the text is designed as an improvement on those treatments intended to capture certain insights that motivated them.
37. One could have versions of these theories in which semantic contents were not objects of the attitudes, but only by foregoing the strong motivation the attitudes provide for these theories.
38. Strictly speaking, the intensions of directly referential terms in (8*b*) and (8*c*) should be constant functions from circumstances to objects, rather than objects themselves. However, this does not affect the issues at hand.
39. The significance of this result is enhanced by the defence, in section V, of the consequence (23) of (A2), (A4), and compositionality. However, it should be noted that analogous results involving the encoding of structure in objects of the attitudes can be established using the weakenings in IV.
40. I leave it open whether there are such expressions.
41. Here I am using '(35*a*)', '(35*c*)', and '(36)' not as names, but as abbreviations for the sentences they normally name.
42. An analogous argument can be constructed regarding assertion.

XII

REFLEXIVITY*

NATHAN SALMON

In 1983 Mark Richard formulated a new and interesting problem for theories of direct reference with regard to propositional-attitude attributions.[1] The problem was later discovered independently by Scott Soames, who recently advanced it[2] as a powerful objection to the theory put forward by Jon Barwise and John Perry in *Situations and Attitudes*.[3] Interestingly, although both Richard and Soames advocate the fundamental assumption on which their philosophical problem arises, they disagree concerning the correct solution to the problem. In this paper I discuss the Richard–Soames problem, as I shall call it, as well as certain related problems and puzzles involving reflexive constructions in propositional–attitude attributions. I will treat these problems by applying ideas I invoked in *Frege's Puzzle*[4] defending a semantic theory that shares certain features with, but differs significantly from, that of Barwise and Perry. Unlike the theory of *Situations and Attitudes*, the theory of *Frege's Puzzle* has the resources without modification to solve the Richard–Soames problem and related problems.

I

In setting out the Richard–Soames problem, we make some important assumptions. First, we make the relatively uncontroversial assumption that a monadic predicate ⌜believes that *S*⌝, where *S* is a declarative sentence, is simply the result of filling the second argument place of the dyadic, fully extensional predicate 'believes' with the term ⌜that *S*⌝. Furthermore, it is assumed that the contribution made by the dyadic predicate 'believes' to securing the information content (with respect to a time *t*) of, or the proposition expressed (with respect to *t*) by, a declarative sentence in which the predicate occurs (outside of the scope of any

Nathan Salmon, 'Reflexivity', *Notre Dame Journal of Formal Logic* 27 (1986), 401–29.

nonextensional devices, such as quotation marks) is a certain binary relation between believers and propositions, the relation of believing–at–t,[5] and that a term of the form ⌜that S⌝ refers (with respect to a possible context of use c) to the information content (with respect to c) of the sentence S itself. More accurately, the following is assumed:

(*B*) A monadic predicate of the form ⌜believes that S⌝, where S is an (open or closed) sentence, correctly applies (with respect to a possible context of use and an assignment of values to individual variables) to all and only those individuals who stand in the binary belief relation (at the time of the context in the possible world of the context) to the information content of, or the proposition expressed by, S (with respect to that context and assignment).

On this assumption, a sentence of the form ⌜a believes that S⌝, where a is any singular term, is true if and only if the referent of a stands in the belief relation to the information content of S. Thesis (*B*) is generally agreed upon by Fregeans and Russellians alike, and is more or less a commonplace in the literature of the theory of meaning, and of the philosophy of semantics generally.

In addition to thesis (*B*), we assume that ordinary proper names, demonstratives, other single-word indexicals (such as 'he'), and other simple (noncompound) singular terms are, in a given possible context of use, Russellian "genuine names in the strict logical sense".[6] Put more fully, we assume the following anti-Fregean thesis as a hypothesis:

(*R*) The contribution made by an ordinary proper name, demonstrative, or other simple singular term to securing the information content of, or the proposition expressed by, declarative sentences (with respect to a given possible context of use) in which the term occurs (outside of the scope of nonextensional operators, such as quotation marks) is just the referent of the term, or the bearer of the name (with respect to that context of use).

In various alternative terminologies, it is assumed that the *interpretation* (Barwise and Perry), or the *Erkenntniswerte* (Frege), or the *content* (David Kaplan), or the *meaning* (Russell), or the *semantic value* (Soames), or the *information value* (myself) of a proper name, demonstrative, or other simple singular term, with respect to a given context, is just its referent.

It is well-known that the thesis that ordinary proper names are Russellian, in this sense, in conjunction with thesis (*B*), gives rise to problems in propositional–attitude attributions, and is consequently relatively unpopular. (Even Russell rejected it.) Thus, thesis (*R*) is hardly the sort of thesis that can legitimately be taken for granted as accepted by the reader. However, I defend thesis (*R*) at some length and in some detail in *Frege's Puzzle*. Moreover, the thesis has gained some long overdue respectability recently, and it cannot be summarily dismissed as obviously misguided. It is (more or less) accepted by Barwise–Perry, Kaplan, Richard, Soames, and others. One standard argument against the thesis—the argument from apparent failure of substitutivity in propositional–attitude contexts—has been shown by Kripke[7] to be inconclusive at best, and the major rival approaches to the semantics of proper names and other simple singular terms have been essentially refuted by Keith Donnellan, Kripke, Perry, and others.[8] The Richard–Soames problem is a problem that arises only on the assumption of thesis (*R*), and it is a problem for this thesis. It is not a problem for alternative approaches, such as those of Frege or Russell, which have much more serious problems of their own. Thesis (*R*) is to be taken as a hypothesis of the present paper, its defence given elsewhere. The conclusions and results reached in the present paper on the assumption of thesis (*R*) may be regarded as having the form "If thesis (*R*) is true, then thus-and-so". The present paper, in combination with *Frege's Puzzle*, allows for the all-important *modus ponens* step.

One version of the Richard–Soames problem can be demonstrated by the following sort of example, derived from Richard's. Suppose that Lois Lane, who is on holiday somewhere in the wilderness, happens to overhear an elaborate plot by some villainous misanthrope to expose Superman to Kryptonite (the only known substance that can harm Superman) at the Metropolis Centennial Parade tomorrow. She quickly rushes for the nearest telephone to warn Superman, but suddenly remembers that the nearest telephone is one day's journey away. As luck would have it, she happens to be standing in front of an overnight mail delivery service outlet. She quickly scribbles a note warning of the plot to harm Superman—a note that absolutely, positively has to get there overnight. She has no address for Superman (or so she believes), but she does have Clark Kent's address, and she (thinks she) knows that Clark planned to spend all day tomorrow at his flat. Now the following sentence is true:

(1*a*) Lois believes that she will directly inform Clark Kent of Superman's danger with her note.

By the assumption of theses (*B*) and (*R*), it would seem that the following sentence contains the very same information as (1*a*), and hence must be true as well:

(1*b*) Lois believes that she will directly inform Superman of Superman's danger with her note.

Richard argues, however, that although (1*a*) is true in this example, (1*b*) cannot be true. For if (1*b*) were true, then the following sentence would also be true:

(1*c*) Lois believes that there is someone *x* such that she will directly inform *x* of *x*'s danger with her note.

That is, if (1*b*) were true, then Lois would also believe that someone or other is such that she will inform him of *his own* danger with her note, since this follows trivially by existential generalization from what she believes according to (1*b*). Yet Lois believes no such thing. (Recall that Lois believes that she has no address for Superman). Of course, Lois hopes that Clark will relay the warning to Superman before it is too late, but she has not formed the opinion that she herself will directly inform someone of his own danger with her note. To put it another way, it is simply false that Lois believes that there is someone with the special property that he will be directly informed by her of his own danger with her note. On the contrary, what she believes is that she will inform someone of *someone else's* danger with her note. Thus (1*a*) is true, though (1*b*) would seem to be false. This poses a serious problem for any theory—such as the theory formed from thesis (*R*) coupled with thesis (*B*) and some other natural assumptions—that claims that (1*a*) and (1*b*) have exactly the same information content, or even merely that they have the same truth value.

Using a similar example, Soames provides a powerful argument against semantic theories of a type that identify the information contents of declarative sentences with sets of *circumstances* (of some sort or other) with respect to which those sentences are either true or untrue (or equivalently, with characteristic functions from circumstances to truth values)—such as the possible-world theories of information content (David Lewis, Robert Stalnaker, and many others) or the "situation" theory of *Situations and Attitudes*. The argument is this: the following sentence concerning a particular ancient astronomer is assumed to be true (where reference to a language, such as 'English', is suppressed):

(2*a*) The astronomer believes: that 'Hesperus' refers to Hesperus and 'Phosphorus' refers to Phosphorus.

Hence according to thesis (*R*) in conjunction with thesis (*B*) and some natural assumptions, the following sentence, which allegedly contains the very same information as (2*a*), must also be true:

(2*b*) The astronomer believes: that 'Hesperus' refers to Hesperus and 'Phosphorus' refers to Hesperus.

But if (2*b*) is true, and thesis (*B*) is also true, then on certain assumptions that are either trivial or fundamental to a set-of-circumstances theory of information content, the following is also true:

(2*c*) The astronomer believes: that something or other is such that 'Hesperus' refers to it and 'Phosphorus' refers to it.

Assuming thesis (*B*), the additional assumptions needed to validate the move from (2*b*) to (2*c*) on any set-of-circumstances theory of information content are: (i) that a believer's beliefs are closed under *simplification* inferences from a conjunction to either of its conjuncts, i.e. if *x* believes *p and q*, then *x* believes *q*; and (ii) that the conjunction of an ordinary sentence *S* (excluding nonreferring singular terms and nonextensional devices such as the predicate 'does not exist') and any existential generalization of *S* is true with respect to exactly the same circumstances as *S* itself.

Now (2*c*) is tantamount to the claim that the astronomer believes that 'Hesperus' and 'Phosphorus' are co-referential. Yet certainly (2*c*) is no consequence of (2*a*). Indeed, we may take it as an additional hypothesis that (2*c*) is false of the ancient astronomer in question. Since (2*a*) is true and (2*c*) is false, it is either false that if (2*a*) then (2*b*)—contrary to the conjunction of theses (*B*) and (*R*)— or else it is false that if (2*b*) then (2*c*)—contrary to the conjunction of (*B*) and any set-of-circumstances theory of information content. Now (*B*) and (*R*) are true. Therefore, Soames argues, any set-of-circumstances theory of information content is incorrect. As Soames points out, the problem points to a fundamental error in the theory of *Situations and Attitudes*, which accepts both (*B*) and (*R*) as fundamental, thereby ensuring the validity of the move from (2*a*) to (2*b*), as well as the assumptions that validate the move from (2*b*) to (2*c*).

In the general case, we may have the first of the following three sentences true and the third false, where *a* and *b* are co-referential proper names, demonstratives, other simple singular terms, or any combination thereof, and *R* is a dyadic predicate:

(3*a*) *c* believes that *aRb*

(3*b*) *c* believes that *aRa*

(3*c*) *c* believes that $(\exists x)xRx$.

The Richard–Soames problem is that (3*b*) appears to follow from (3*a*), and (3*c*) appears to follow from (3*b*). Since (3*a*) is true and (3*c*) false, something has got to give.

II

Now (3*b*) is either true or false. Hence it is either false that if (3*a*) then (3*b*), or else it is false that if (3*b*) then (3*c*). Both Richard and Soames accept thesis (*R*). Insisting that if (3*b*) then (3*c*), Richard maintains that it is false that if (3*a*) then (3*b*), thereby impugning thesis (*B*).[9] Accepting thesis (*B*) as well as (*R*), Soames argues instead that "there is a principled means of blocking" the move from (3*b*) to (3*c*) while preserving (*B*).

There is a certain intuitive picture of belief advanced by Barwise and Perry (Chapter 10) and which is independently plausible in its own right. This is a picture of belief as a cognitive state arising from internal mental states that derive information content in part from causal relations to external objects. Soames points out that on this picture of belief, the following is indeed true if (3*b*) is:

(3*d*) $(\exists x)$ *c* believes that *xRx*.

Soames adds:[10]

However, [on this picture of belief] there is no reason to think that [the referent of *c*] believes the proposition that something bears *R* to itself. Since none of the agent's mental states has this as its information content, he does not believe it.

Quine distinguishes two readings of any sentence of the form ⌜c believes something is ϕ⌝—what he calls the *notional* and the *relational* readings. The notional reading may be spelt out as ⌜c believes: that something or other is ϕ⌝. It is the Russellian secondary occurrence or small–scope reading. The relational reading may be spelt out as ⌜c believes something in particular to be ϕ⌝, or more perspicuously as ⌜Something is such that c believes: that it is ϕ⌝. It is the Russellian primary occurrence or large–scope reading. In Quine's terminology, Soames claims that the notional reading of ⌜c believes something bears R to it⌝ does not follow from the relational. Quine demonstrated some time ago that the relational reading of ⌜c believes something is ϕ⌝ does not in general follow from the notional reading, with his clever example of 'Ralph believes someone is a spy'. Soames may be seen as arguing that, on a certain plausible picture of belief, there are cases in which the reverse inference also fails. Since the appearance of Quine's influential writings on the subject, it is no longer surprising that the notional reading does not imply the relational. It is at least somewhat surprising, however, that there could be converse cases in which the relational reading is true yet the notional reading false. This is what Soames is arguing.

My own view of the Richard–Soames problem favours Soames's account over Richard's. Thesis (B) is supported by strong linguistic evidence. It provides the simplest and most plausible explanation, for example, of the validity of such inferences as:

> John believes the proposition to which our nation is dedicated.
> Our nation is dedicated to the proposition that all men are created equal.
> Therefore, John believes that all men are created equal.

Furthermore, although a number of philosophers have proposed a variety of truth-condition assignments for belief attributions contrary to thesis (B), these alternative truth-condition assignments often falter with respect to belief attributions that involve open sentences as their complement 'that'-clause, and that are true under some particular assignment of values to individual variables or to pronouns—for example, 'the astronomer believes that x is a planet' in 'There is something x such that x = Venus and the astronomer believes that x is a planet' or 'the astronomer believes that it is a planet' in 'As regards Venus, the astronomer believes that it is a planet'.[11] Thesis (B) should be maintained to the extent that the facts allow, and should not be abandoned if Soames is correct that there is a principled means of solving the Richard–Soames problem while maintaining (*B*).

By contrast, Soames's proposals for solving the problem invoke essentially some of the same ideas advanced and defended in *Frege's Puzzle*. There I develop and defend thesis (*R*) (and, to a lesser extent, thesis (*B*)), as well as the view (which Russell himself came to reject) that the contents of beliefs formulatable using ordinary proper names, demonstratives, or other simple singular terms, are so-called singular propositions (Kaplan), i.e. structured propositions directly about some individual which occurs directly as a constituent of the proposition. I take propositions to be structured in such a way that the structure and constituents of a proposition are directly readable from the structure and constituents of a declarative sentence containing the proposition as its information content. By and large, a simple (noncompound) expression contributes a single entity, taken as a simple (noncomplex) unit, to the information content of a sentence in which the expression occurs, whereas the contribution of a compound expression (such as a phrase or sentential component) is a complex entity composed of the contributions of the simple components.[12] One consequence of this sort of theory is that, contrary to set-of-circumstances theories of information content, there is a difference, and therefore a distinction, between the information content of the conjunction of an ordinary sentence *S* and any of its existential generalizations and that of *S* itself. This disables the argument that applied in the case of a set-of-circumstances theory to establish the (alleged) validity of the move from (3*b*) to (3*c*).

Unfortunately, this difference between the two sorts of theories of information content does not make the problem disappear altogether. There is an interesting philosophical puzzle concerning the logic and semantics of propositional–attitude attributions that is generated by the Richard–Soames problem, a puzzle that arises even on the structured-singular-proposition sort of view sketched above.

Soames slightly misstates the case when he says that (on the intuitive picture of belief as deriving from certain mental states having information content), "there is no reason to think that (3*c*) is true". For in fact, even though (1*c*) and (2*c*) are false in the above examples, there are very good reasons to think that they are true. One excellent reason to think that (1*c*) is true is the fact that

(1*b*) is true, and one excellent reason to think that (2*c*) is true is the fact that (2*b*) is true. In general, it is to be expected that if a sentence of the form ⌜*c* believes that ϕ_a⌝ is true, then so is ⌜*c* believes that $(\exists x)\phi_x$⌝, where *a* is a singular term that refers to something, ϕ is an ordinary extensional context (excluding predicates such as 'does not exist'), and ϕ_a is the result of substituting (free) occurrences of *a* for free occurrences of '*x*' uniformly throughout ϕ_x. There is a general psychological law to the effect that subjects typically tend to believe the existential generalizations of their beliefs. Herein the puzzle arises. Even if the conjunctive proposition *'Hesperus' and 'Phosphorus' refer to Hesperus and there is something that 'Hesperus' and 'Phosphorus' refer to* is not the same proposition as the simpler proposition *'Hesperus' and 'Phosphorus' refer to Hesperus*, if the astronomer believes that 'Hesperus' and 'Phosphorus' refer to Hesperus, then it seems he ought to believe that there is something that 'Hesperus' and 'Phosphorus' refer to. And if Lois believes that she will inform Superman of his danger with her note, then it seems she ought to believe that there is someone whom she will inform of his danger with her note. It is precisely for this reason that Richard rejects (1*b*), even though he does not endorse a set-of-circumstances theory of information content and favours the structured-singular-proposition account.

Perhaps if a subject is insane or otherwise severely mentally defective, he or she may fail to believe the (validly derivable) existential generalizations of his or her beliefs, but we may suppose that neither Lois Lane nor the astronomer suffer from any mental defects. We may even suppose that they are master logicians, or worse yet, that they have a perverse penchant for drawing existential generalization (EG) inferences as often as possible. They go around saying things like 'I'm tired now; hence, sometimes someone or other is tired' and 'Fred shaves Fred; hence someone shaves Fred, Fred shaves someone, and someone shaves himself'. In this way, it can be built into the example that the truth of (1*b*) is an excellent reason to believe in the truth of (1*c*), and the truth of (2*b*) is an excellent reason to believe in the truth of (2*c*). For such EG-maniacs, one might expect that it is something of a general law that every instance of the following schema is true:

(L_1) If *c* believes that ϕ_a, then *c* believes that something is such that ϕ_{it}, where c refers to the subject, a is any referring singular term of English, ϕ_{it} is any English sentence in which the pronoun 'it' occurs (free and not in the scope of quotation marks, an existence predicate, or other such operators) and which may also contain occurrences of a, and ϕ_a is the result of substituting (free) occurrences of a for (free) occurrences of 'it' throughout ϕ_{it}. In fact, one might expect that it is something of a general law that every instance of (L_1) is true where c refers to any normal speaker of English, even if he or she is not an EG-maniac.

I maintain with Soames that the sentences ⌜If (1*b*) then (1*c*)⌝ and ⌜If (2*b*) then (2*c*)⌝ constitute genuine counterexamples to this alleged general law. But even if the principle that every instance of (L_1), as formulated, is true is thereby refuted, surely something very much like it, some weakened version of it, *must be* true—even where the referent of c does not have a perverse penchant for existential generalization. For the most part, in the typical kind of case, it would be highly irrational for someone to fail to believe the existential generalizations of one of his or her beliefs. Neither Lois Lane nor the astronomer is irrational in this way. The conditionals ⌜If (1*b*) then (1*c*)⌝ and ⌜If (2*b*) then (2*c*)⌝ are not typical instances of schema (L_1), but it is not enough simply to point out how they are atypical and to leave the matter at that. It is incumbent on the philosopher who claims that these instances of (L_1) fail, to offer some alternative principle that is *not* falsified in these cases and thereby accounts for the defeasible reliability, and the *prima facie* plausibility, of the alleged general law.

This is not a problem special to set-of-circumstances theories of information content. It is equally a puzzle for the structured-singular-proposition sort of theory that I advocate and that Soames proposes in his discussion of the Richard–Soames problem. It is a puzzle for the conjunction of theses (*B*) and (*R*), irrespective of how these theses are supplemented with a theory of information content.

III

There is a second, and surprisingly strong, reason to suppose that (1*c*) and (2*c*) are true. The general puzzle posed by the Richard–Soames problem can be significantly strengthened if we exploit a simple reflexive device already present to a certain degree in standard English.

Given any simple dyadic predicate Π, we may form a monadic pedicate ⌜*self*-Π⌝ defined by

$(\lambda x)x\Pi x$,

in such a way that ⌜*self*-Π⌝ is to be regarded as a simple (noncompound) expression, a single word. In English, this might be accomplished by converting a present tensed transitive verb *V* into a corresponding adjective and prefixing 'self-' to obtain a reflexive adjective; e.g. from 'cleans' we obtain 'self-cleaning', from 'indulges', 'self-indulgent', from 'explains', 'self-explanatory', and so on. The contribution made by a term of the form ⌜*self*-Π⌝ to the information content, with respect to a time *t*, of a typical sentence in which it occurs is simply the reflexive property of bearing *R* to oneself at *t*, where *R* is the binary relation semantically associated with Π.[13] Assuming thesis (*R*), if *a* is a proper name or other simple singular term and *R* is the binary relation semantically associated with Π, then the information content, with respect to *t*, of the sentence ⌜*self*-Π(*a*)⌝ is the singular proposition made up of the referent of *a* together with the property of bearing *R* to oneself at *t*.

Consider again the move from (3*a*) to (3*b*), where *a* and *b* are co-referential proper names, *R* is a simple dyadic predicate, and (3*a*) is true:

(3*a*) *c* believes that *aRb*

(3*b*) *c* believes that *aRa*.

As Soames points out, (on a plausible picture of belief) the following relational, or *de re*, attribution follows from (3*b*):

(3*d*) $(\exists x)$ *c* believes that *xRx*.

In fact, a somewhat stronger *de re* attribution also follows from (3*b*), by exportation:[14]

$(\exists x)$ [$x = a$ & *c* believes that *xRx*],

or less formally:

(3*b*′) *c* believes of *a* that it *R* it.

Now from this it would seem to follow that:

(3*e*′) *c* believes of *a* that it *R* itself.

From this (perhaps together with some general psychological law) it would seem to follow further that:

(3*f*′) *c* believes of *a* that *self-R*(it),

with the predicate ⌜*self-R*⌝ understood as explained above. Finally by importation, we may infer:

(3*f*) *c* believes that *self-R*(*a*).

For example, suppose that, owing to certain miscalculations, the astronomer comes to believe that Hesperus weighs at least one thousand tons more than Phosphorus. Now every step in the following derivation follows by an inference pattern that is either at least apparently intuitively valid or else sanctioned by the conjunction of theses (*B*) and (*R*), or both:

(4*a*) The astronomer believes that Hesperus outweighs Phosphorus.

(4*b*) The astronomer believes that Hesperus outweighs Hesperus.

(4*b*′) The astronomer believes of Hesperus that it outweighs it.

(4*e*′) The astronomer believes of Hesperus that it outweighs itself.

(4*f*′) The astronomer believes of Hesperus that it is self-outweighing.

(4*f*) The astronomer believes that Hesperus is self-outweighing.

One could continue the sequence of inferences from (4*f*) all the way to:

(4*c*) The astronomer believes that there is something such that it outweighs it.

by invoking some corrected, weakened version of the law mentioned above (the alleged law that every appropriate instance of (L_1) is true), to pass from (4*f*) to:

(4*g*) The astronomer believes that there is something such that it is self-outweighing,

from which (4*c*) appears to follow directly. But there is no need to extend the derivation this far. A problem arises at least as soon as (4*f*). For unless the astronomer is insane, or otherwise severely mentally defective, (4*f*) is obviously false. The astronomer would not ascribe to Venus the reflexive property, which nothing could possibly have, of weighing more than oneself. Hence, in moving from a sentence to its immediate successor, somewhere in the derivation of (4*f*) we move from a truth to a falsehood. Where? The moves from (4*a*) to (4*b*′) and from (4*f*′) to (4*f*) are validated by the conjunction of theses (*B*) and (*R*), and both of the remaining transitions commencing with (4*b*′) are based on inference patterns that (assuming ordinary folk psychology and that the astronomer is normal) seem intuitively valid.

One may harbour some residual doubts about the exportation move from (4*b*) to (4*b*′) and/or the importation move from (4*f*′) to (4*f*). The theory formed from the conjunction of theses (*B*) and (*R*) requires the validity of both of these inferences, so that if either is invalid the theory is false. In fact, however, these inferences are not essential to the present puzzle. The exportation inference takes us on a detour that some may find helpful, though one may bypass the *de re* 'believes of' construction altogether. Instead, we may construct the following alternative derivation from (4*b*):

(4*b*) The astronomer believes that Hesperus outweighs Hesperus.
(4*e*) The astronomer believes that Hesperus outweighs itself.
(4*f*) The astronomer believes that Hesperus is self-outweighing.

If the inference from (4*b*′) to (4*e*′) is valid, then by parity of reasoning so is the inference from (4*b*) to (4*e*). And if the inference from (4*e*′) to (4*f*′) is valid, then by parity of reasoning so is the inference from (4*e*) to (4*f*). Hence, if the derivation of (4*f*′) from (4*b*′) via (4*e*′) is legitimate, then so is the derivation of (4*f*) from (4*b*) via (4*e*). But (4*f*) is false. Therefore, it would seem, so is (4*b*). Sentence (4*a*), on the other hand, is true. This raises anew doubts about the independently suspicious move from (4*a*) to (4*b*), or more generally, the move from (3*a*) to (3*b*), thereby impugning once again the conjunction of theses (*R*) and (*B*).

The new puzzle, then, is this: according to the conjunction of theses (*B*) and (*R*), (4*b*) follows from (4*a*) together with the fact that 'Hesperus' and 'Phosphorus' are co-referential proper names. Now in the sequence ⟨(4*b*), (4*e*), (4*f*)⟩, each sentence appears to follow logically from its immediate predecessor. Alternatively in the sequence ⟨(4*b*′), (4*e*′), (4*f*′)⟩, each sentence appears to follow logically from its immediate predecessor, and furthermore, according to the conjunction of (*B*) and (*R*), (4*b*) entails (4*b*′), and

(4*f*′) entails (4*f*). One way or another, we seem to be able to derive (4*f*) from (4*a*), together with the fact that 'Hesperus' and 'Phosphorus' are co-referential proper names. Yet in the example, (4*a*) is plainly true and (4*f*) plainly false. Where does the derivation go wrong?

I call this *the puzzle of reflexives in propositional attitudes*. Here again, the problem posed by the puzzle is especially pressing for any set-of-circumstances theory of information content. In fact, the problem is even more pressing than the Richard–Soames problem for such theories, if that is possible. One difference between the Richard–Soames problem and the puzzle of reflexives in propositional attitudes is that what is said to be believed at the final step of the derivation, in this case step (3*f*), is not merely a consequence of, but is *equivalent to*, what is said to be believed in (3*b*). In fact, any circumstance in which an individual x bears R to x is a circumstance in which x has the reflexive property of bearing R to oneself, and vice versa. There is no need here to make the additional assumption that belief is closed under simplification inferences. Any set-of-circumstances theory of information content, in conjunction with thesis (*B*), automatically validates the derivation of (3*f*) from (3*b*). The problem thus also points to a fundamental error in the theory of *Situations and Attitudes* which includes both theses (*B*) and (*R*) as fundamental, thereby validating the full derivation of (3*f*) from (3*a*) without any further assumptions concerning belief. The puzzle of reflexives in propositional attitudes, however, is not peculiar to set-of-circumstances theories, and arises on any theory of information content that incorporates the conjunction of theses (*B*) and (*R*), including the structured-singular-proposition theory that I advocate. The difference is that the structured-singular-proposition view (in conjunction with (*B*) and (*R*)), unlike the theory of *Situations and Attitudes*, is not committed by its very nature to the validity of the derivation of (4*f*) from (4*b*). It is just that each step in the derivation of (4*f*) from (4*b*) is independently plausible.

IV

The puzzle of reflexives in propositional attitudes is related to a paradox that concerns quantification into belief contexts and that was discovered some time ago by Alonzo Church.[15] Unlike the former puzzle, however, Church's paradox presents a serious problem in particular for the theory of structured singular propositions.

As a matter of historical fact, as of some appropriate date, King George IV was acquainted with Sir Walter Scott, but was doubtful whether Scott was the author of *Waverley*. We may even suppose that George IV believed at that time that Scott did not write *Waverley*. Yet, Church notes, if quantification into belief contexts is taken as meaningful in combination with the usual laws of the logic of quantification and identity, then the following is provable as a logical theorem using classical Indiscernibility of Identicals (Leibniz's Law):

(5) For every x and every y, if Georgy IV does not believe that $x \neq x$, if George IV believes that $x \neq y$, then $x \neq y$.

Mimicking the standard proof in quantified modal logic of the necessity of identity, Church remarks that although it is not certain, it was very likely true as of the same date that:

(6) For every x, George IV does not believe that $x \neq x$,

since it is very likely that George IV did not believe anything to be distinct from itself. Taking (6) as premiss, we may derive:

(7) For every x and every y, if George IV belives that $x \neq y$, then $x \neq y$.

We are thus apparently led to ascribe to King George's beliefs the strange "power to control the actual facts about x and y". Since Scott is in fact the author of *Waverley*, this derivation of (7) from (6) seems to preclude King George's believing, as of the same date, that Scott did not write *Waverley*. The derivation thus constitutes an unacceptable paradox, not unlike Russell's paradox of naïve set theory (set theory with unrestricted comprehension). Church concludes that this provides a compelling reason to reject the meaningfulness of quantification into belief contexts.[16]

As quantification into belief contexts goes, so goes the theory of structured singular propositions as potential objects of belief. Church's paradox thus poses a serious difficulty for the theory that I advocate. But it also poses a serious difficulty for any theory, including any set-of-circumstances theory, that purports to make sense of *de re* constructions or quantification into belief contexts. Furthermore, the paradox is quite independent of the conjunction of theses (*B*) and (*R*). Whether these are true or false, the paradox arises as long as quantification into belief contexts is regarded as meaningful.

V

It is precisely to treat philosophical puzzles and problems of the sort presented here that I proposed the sketch of an analysis of the binary belief relation between believers and propositions (sometimes Russellian singular propositions) in *Frege's Puzzle*. I take the belief relation to be, in effect, the existential generalization of a ternary relation, *BEL*, among believers, propositions, and some third type of entity. To believe a proposition *p* is to adopt an appropriate favourable attitude toward *p* when taking *p* in some relevant *way*. It is to agree to *p*, or to assent mentally to *p*, or to approve of *p*, or some such thing, when taking *p* a certain way. This is the *BEL* relation. The third relata for the *BEL* relation are something like *proposition guises*, or *modes* of acquaintance with propositions, or *ways* in which a believer may be familiar with a given proposition. Of course, to use a distinction of Kripke's, this formulation is far too vague to constitute a fully developed *theory* of belief, but it does provide a *picture* of belief that differs significantly from the sort of picture of propositional attitudes advanced by Frege or Russell, and enough can be said concerning the *BEL* relation to allow for at least the sketch of a solution to certain philosophical puzzles, including the original puzzle generated by the Richard–Soames problem.

In particular, the *BEL* relation satisfies the following three conditions:

(i) *A* believes *p* if and only if there is some *x* such that *A* is familiar with *p* by means of *x* and *BEL*(*A*, *p*, *x*).

(ii) *A* may believe *p* by standing in *BEL* to *p* and some *x* by means of which *A* is familiar with *p* without standing in *BEL* to *p* and all *x* by means of which *A* is familiar with *p*.

(iii) In one sense of 'withhold belief', *A* withholds belief concerning *p* (either by disbelieving or by suspending judgement) if and only if there is some *x* by means of which *A* is familiar with *p* and not-*BEL*(*A*, *p*, *x*).

These conditions generate a philosophically important distinction between withholding belief and failure to believe (i.e. not believing). In particular, one may both withhold belief from and believe the very same proposition simultaneously. (Neither withholding belief nor failure to believe is to be identified with the related notions of disbelief and suspension of judgement—which are two different ways of withholding belief, in this sense, and which may occur simultaneously with belief of the very same proposition in a single believer.)

It happens in most cases (but not all) that when a believer believes some particular proposition *p*, the relevant third relatum for the *BEL* relation is a function of the believer and some particular *sentence* of the believer's language. Consider for example the binary function *f* that assigns to any believer *A* and sentence *S* of *A*'s language, the *way A* takes the proposition contained in *S* (in *A*'s language with respect to *A*'s context at some particular time *t*) were it presented to *A* (at *t*) through the very sentence *S*. Then (assuming *t* is the time in question) Lois believes the proposition that she will inform Clark Kent of Superman's danger with her note by virtue of standing in the *BEL* relation to this proposition together with the result of applying the function *f* to Lois and the particular sentence 'I will inform Clark Kent of Superman's danger with my note'. That is, in the example the following is true:

> *BEL*(Lois, that she will inform Clark Kent of Superman's danger with her note, *f*[Lois, 'I will inform Clark Kent of Superman's danger with my note']).

On the other hand, the following is false:

> *BEL*(Lois, that she will inform Superman of his danger with her note, *f* [Lois, 'I will inform Superman of his danger with my note']).

Similarly, assuming the astronomer in Soames's example spoke English:

> *BEL*(the astronomer, that 'Hesperus' refers to Hesperus and 'Phosphorus' refers to Phosphorus, *f* [the astronomer, "'Hesperus' refers to Hesperus whereas 'Phosphorus' refers to Phosphorus"]),

but not:

> *BEL*(the astronomer, that 'Hesperus' refers to Hesperus and

> 'Phosphorus' refers to Hesperus, *f* [the astronomer, " 'Hesperus' and 'Phosphorus' both refer to Hesperus"]).

In *Frege's Puzzle* the *BEL* relation and the function *f* are invoked in various ways to explain and to solve some of the standard (and some nonstandard) problems that arise on the sort of theory I advocate. This device is also useful with regard to the original puzzle that arises from the Richard–Soames problem and the puzzle of reflexives in propositional attitudes.

In the first example, (1*c*) is false, since Lois does not adopt an appropriate favourable attitude toward the proposition that there is someone whom she will inform of his own danger with her note, no matter how this proposition might be presented to her. That is, there is no *x* such that Lois stands in *BEL* to the proposition that she will inform someone or other of his own danger with her note and *x*. Similarly, in Soames's example. (2*c*) if false, since the astronomer does not adopt the appropriate favourable attitude toward the proposition that 'Hesperus' and 'Phosphorus' are co-referential, no matter how this proposition might be presented to him. He does not stand in *BEL* to this proposition and any *x*.

What about (1*b*) and (2*b*)? These are indeed true in the examples. Consider the first example. Sentence (1*a*) is true by hypothesis. Now notice that if Superman were somehow made aware of the truth of (1*a*), then he could truthfully utter the following sentence:

(1*b*I) Lois believes that she will directly inform me of my danger with her note.

In fact, (1*b*I) yields the only natural way for Superman to express (to himself) the very information that is contained in (1*a*). But if (1*b*I) is true with respect to Superman's context, then (1*b*) is true with respect to ours. Both (1*b*I), taken with respect to Superman's context, and (1*b*), taken with respect to ours, are true precisely because Lois adopts the appropriate favourable attitude toward the proposition about Superman, i.e. Clark Kent, that she will inform him of his danger with her note. Lois assents to this information when she takes it the way she would if it were presented to her through the sentence 'I will inform Clark Kent of Superman's danger with my note'. Hence, she believes it. Similarly, the astronomer inwardly assents to the proposition about Hesperus, i.e. Venus, that 'Hesperus' refers to it and

'Phosphorus' refers to it, when it is presented to him through the sentence " 'Hesperus' refers to Hesperus whereas 'Phosphorus' refers to Phosphorus". Hence (2*b*) is true.

In fact, in the examples Lois also believes that she will *not* inform Superman of his danger with her note, and the astronomer that 'Hesperus' and 'Phosphorus' do *not* both refer to Hesperus, since:

> *BEL*(Lois, that she will not inform Superman of his danger with her note, *f* [Lois, 'I will not inform Superman of his danger with my note'])

and:

> *BEL*(the astronomer, that 'Hesperus' and 'Phosphorus' do not both refer to Hesperus, *f* [the astronomer, " 'Hesperus' and 'Phosphorus' do not both refer to Hesperus"]).

Both Lois and the astronomer thus (unknowingly) believe some proposition together with its denial.[17]

One reason so many instances of schema (L_1) are true, although it fails in these special cases, is that the schema approximates the following weaker schema, all (or at least very nearly all) of whose instances are true, and which is not falsified in these special cases:

> (L_2) If $(\exists p)BEL(c, p, f[c,$ 'ϕ_a'$])$, then $(\exists q)BEL(c, q, f[c,$ 'Something is such that ϕ_{it}'$])$,

where c refers to a normal speaker of English, a is any referring singular term of English, ϕ_{it} is any English sentence in which the pronoun 'it' occurs (free and not in the scope of quotation marks, an existence predicate, or other such operators) and which may also contain occurrences of a, and ϕ_a is the result of substituting (free) occurrences of a for (free) occurrences of 'it' throughout ϕ_{it}. I submit that the similarity of the former schema (L_1) to something like schema (L_2) is a major source of the plausibility of the alleged general law concerning the former. Schema (L_2) is not falsified in these special cases, even if Lois and the astronomer are normal speakers of English, since Lois does not agree to the proposition that she will inform Superman of his danger with her note when she takes it in the way she would if it were presented to her through the sentence 'I will inform Superman of his danger with my note', and the astronomer does not agree to the proposition that 'Hesperus' and 'Phosphorus' refer to Venus when it is presented to him through the sentence " 'Hesperus' and 'Phosphorus' refer to Hesperus".[18]

VI

Even if this resolves the original puzzle generated by the Richard–Soames problem for the structured-singular-proposition account of information content, it does not yet lay to rest the puzzle of reflexives in propositional attitudes, not to mention Church's ingenious paradox concerning quantification into belief contexts.

Richard's proposal to solve the original puzzle by blocking the initial inference from (3*a*) (together with the fact that *a* and *b* are co-referential proper names or other simple singular terms) to (3*b*) would equally block the puzzle of reflexives in propositional attitudes. This proposal involves relinquishing thesis (*B*), and is motivated by the threat of the alleged derivability of falsehoods such as (1*c*) from (1*b*). But I argued above that thesis (*B*) is supported by strong linguistic evidence, and should be maintained insofar as the facts allow. We have seen that the account of belief in terms of the *BEL* relation effectively blocks the move from (1*b*) to (1*c*), while retaining thesis (*B*) and while also affording an explanation (or at least the sketch of an explanation) for the *prima facie* plausiblity of the move. If there is a solution to the problem of reflexives in propositional attitudes, it does not lie in the rejection of thesis (*B*).

Ruth Barcan Marcus has argued that, in at least one ordinary sense of 'believe', it is impossible to believe what is impossible.[19] Marcus would thus claim that (4*a*) is false to begin with, since the astronomer cannot "enter into the belief relation" to the information, which is necessarily misinformation, that Hesperus outweighs Phosphorus. However, one of Marcus's arguments for this, perhaps her main argument, appears to be that where *a* and *b* are co-referential names, if (3*a*) is true so is (3*c*), and in a great many cases where one is inclined to hold an instance of (3*a*) true even through ⌜*aRb*⌝ encodes necessarily false information ((4*a*) for example), (3*c*) is patently false, because ⌜$(\exists x)xRx$⌝ (e.g. 'Something outweighs itself') encodes information that is not only impossible but patently unbelievable.[20]

Marcus's view that one cannot believe what cannot be true is highly implausible, and I believe, idiosyncratic. It often happens in mathematics and logic that owing to some fallacious argument, one comes to embrace a fully grasped proposition that is in fact provably false. Sometimes this happens even in philosophy, more often than we care to admit. In our example, we may suppose that, for some particular number *n*, the astronomer comes to believe the proposition that Hesperus weighs at least *n* tons, and also the proposition that Phosphorus weighs no more than (*n* – 1,000) tons. He embraces these two propositions. It is very implausible to suppose that the fact that their conjunction is such that it could not be true somehow prevents the astronomer from embracing that conjunction, along with its component conjuncts, or that the astronomer is somehow prevented from forming beliefs on the basis of inference from his two beliefs, as in (4*a*).

More important for our present purpose is that Marcus's argument for the falsehood of (4*a*), at least as the argument is interpreted here, has to be mistaken. Otherwise, one could also show that (1*a*) and (2*a*) are false in the original examples. For although the proposition that Lois will inform someone of his own danger with her note is not unbelievable, it is plain in the example that it is not believed by Lois, i.e. (1*c*) is plainly false. If Marcus's argument for the impossibility of believing the impossible were sound, then by parity of reasoning it would follow that (1*a*) is false. Similarly, although the proposition that 'Hesperus' and 'Phosphorus' are co-referential is believable, in fact true, it is a hypothesis of the example that the astronomer does not believe it, i.e. (2*c*) is stipulated to be false. If Marcus's claim that (3*c*) is true if (3*a*) is true were itself true, it would follow that (2*a*) is false. But (1*a*) and (2*a*) are plainly true in these examples. There must be something wrong, therefore, with Marcus's argument, at least as I have interpreted it here.

What is wrong is precisely the claim that (3*c*) is true if (3*a*) is. Since it is incorrect, this claim cannot give us a way out of the present problem. In fact by shifting from (4*a*)–(4*f*) to another example, we can remove the feature that what is said to be believed at step (*a*) is such that it could not be true. Thus from 'Lois believes that Clark Kent disparages Superman while Superman indulges Clark Kent' we may construct a parallel and equally fallacious derivation of 'Lois believes that Superman is self-disparaging and self-indulgent'. Marcus's unusual contention that it is impossible to believe the impossible, whether correct or incorrect, is simply irrelevant to this example.

What, then, is the solution to the puzzle of reflexives in propositional attitudes for the theory of structured singular propositions?

In the example, (4*b*) and (4*b*′) are true, whereas (4*f*) and (4*f*′) are false. Any temptation to infer (4*f*) from (4*b*), or (4*f*′) (4*b*′), can be explained using the *BEL* relation and the function *f* in a manner similar to the explanation given above in connection with the *prima facie* plausibility of inferring (3*c*) from (3*b*). In any case, either the inference from (4*b*) to (4*e*) (and therewith the inference from (4*b*′) to (4*e*′)) is fallacious, or the inference from (4*e*) to (4*f*) (and therewith the inference from (4*e*′) to (4*f*′)) is. Which is it?

Answering this question involves taking sides in a current controversy concerning the identity or distinctness of propositions of the form *x bears R to x* and *x bears R to itself*. If the propositions that Hesperus outweighs Hesperus and that Hesperus outweighs itself are the very same, then the inference from (4*b*) to (4*e*) is valid by classical Indiscernibility of Identicals (or Leibniz's Law) together with thesis (*B*), and the inference from (4*e*) to (4*f*) must then be rejected. If, on the other hand, these propositions are not the same and instead the proposition that Hesperus outweighs itself is the same (or very nearly the same) as the proposition that Hesperus is self-outweighing, then the inference from (4*e*) to (4*f*) is unobjectionable and the inference from (4*b*) to (4*e*) must be rejected.

As noted in Section III above, the advocate of a set-of-circumstances theory of information content is committed to the claim that propositions of the form *x bears R to x* and *x bears R to itself* are exactly the same, since any circumstance in which *x* bears *R* to *x* is one in which *x* bears *R* to itself, and vice versa. Thus, M. J. Cresswell, a set-of-possible-worlds theorist, has recently claimed that:[21]

> on any reasonable account of propositions, the proposition that Ortcutt loves himself ought to be the same as the proposition that Ortcutt loves Ortcutt.

This, however, is far from the truth. In fact, there are compelling reasons to distinguish a proposition of the form *x bears R to x* from the proposition *x bears R to itself*. One sort of consideration is the following: we must distinguish between the reflexive property of exceeding oneself in weight and the simple relational property of exceeding the planet Venus in weight. The former is an impossible property; it is quite impossible for anything to possess it. The latter property, on the other hand, is fairly widespread; a great many massive objects (e.g. the stars) possess it—although, of course, it is quite impossible for Venus to possess it. Now the sentence 'Hesperus outweighs itself' seems to ascribe to Hesperus, i.e. Venus, the impossible property of weighing more than oneself, rather than the simple relational property of weighing more than Venus. It seems to say about Venus what 'Mars outweighs itself' says about Mars—that it has the reflexive property of exceeding oneself in weight—and not what 'Mars outweighs Venus' says about Mars. If one wants to ascribe to Venus the simple relational property of weighing more than Venus, rather than the impossible property of weighing more than oneself, one may use the sentence 'Hesperus outweighs Hesperus' (among others). It says about Venus what 'Mars outweighs Venus' says about Mars—that it weighs more than Venus—instead of what 'Mars outweighs itself' seems to say about Mars. If one prefers, it ascribes the relation of exceeding-in-weight to the ordered pair of Venus and itself. In either case, the proposition contained in 'Hesperus outweighs Hesperus' is not the same as what seems to be the proposition contained in 'Hesperus outweighs itself'.[22] Contrary to any set-of-circumstances account of propositions, the proposition about Venus, that it weighs more than it, is a different proposition from the proposition about Venus that it is self-outweighing, although they are, in some sense, logically equivalent to one another.[23] The astronomer in the example believes the former and not the latter. Neither the sentence 'Hesperus outweighs Hesperus' nor the sentence 'Hesperus outweighs itself' can be regarded as somehow containing *both* of these propositions simultaneously (as might be said, for example, of the conjunction 'Venus has the simple relational property of weighing more than Venus and also the reflexive property of weighing more than oneself'). Each sentence contains precisely one piece of information, not two. Neither is ambiguous; neither is a conjunction of two sentences with different (albeit equivalent) information contents.[24] Similar remarks may be made in connection with Cresswell's example of 'Ortcutt loves Ortcutt' and 'Ortcutt loves himself'.

This conception of reflexive propositions of the form *x bears R to itself* involves rejecting the otherwise plausible view that the reflexive pronoun 'itself' in 'Hesperus outweighs itself' refers anaphorically to the planet Venus. Instead, the pronoun might be regarded as a predicate-operator, one that attaches to a dyadic predicate to form a compound monadic predicate. Formally, this operator may be defined by the following expression:[25]

$$(\lambda R)(\lambda x)xRx.$$

The alternative conception of propositions of the form *x bears R to itself* involves treating reflexive pronouns instead as anaphorically referring singular terms. On this view, in order to ascribe to Venus the reflexive impossible property of weighing more than oneself, it is not sufficient to use the sentence 'Hesperus outweighs itself'. Instead, one must resort to some device such as the predicate 'is self-outweighing'.

There can be no serious question about the possibility of an operator such as the one defined above. The displayed expression definitely captures a *possible* operator on dyadic predicates. There is no reason why English (and other natural languages) could not contain such an operator, and there is no *a priori* argument that standard English does not have this operator. The question is whether the reflexive pronouns of standard English ('itself', 'himself', 'myself', 'oneself', etc.) are expressions for this operator, rather than anaphorically referring singular terms.

This is not a metaphysical question about the essential natures of propositions, but an empirical question about the accidents of standard English semantics. It is a question, moreover, for which decisive linguistic evidence is difficult to produce, since on either hypothesis the information content of 'Hesperus outweighs itself' is logically equivalent to the content yielded by the rival hypothesis (although writers on both sides of this dispute have advanced what they take to be compelling evidence for their view).

Assuming that the semantic analysis presented above of sentences such as 'Hesperus outweighs Hesperus' is at least roughly correct, the claim that propositions of the form *x bears R to x* and *x bears R to itself* are the same is tantamount to the empirical claim that the reflexive pronouns of standard English are singular terms and not expressions for the predicate-operator defined above, whereas the claim that the proposition *x bears R to itself* is not the same as *x bears R to x* but instead goes with *x is self-R* is tantamount to the empirical claim that the reflexive pronouns are expressions for the predicate-operator and not singular terms. This issue cannot be settled by *a priori* philosophical theorizing about the nature of propositions. A complete solution to the puzzle of reflexives in propositional attitudes thus turns on answering a difficult empirical question concerning the meanings of reflexive pronouns in standard English.

VII

The time has come to face the music. How can the theory of structured singular propositions solve Church's paradox concerning quantification into belief contexts?

Fortunately, some of the ideas discussed in the preceding sections bear directly on Church's paradox. Notice first that (7), taken literally, does not ascribe any power to King George or his beliefs *per se*. Nor does it ascribe to George an infallibility concerning the distinctness of distinct individuals x and y. It merely states a generalization concerning every pair of individuals x and y believed distinct by King George. In Humean terminology, it merely states a *constant conjunction* between any pair of individuals being believed distinct by King George and their actually being distinct. As Hume noted, there is no idea of power contained in that of constant conjunction. Analogously, the sentence 'All crows are black' merely states a generalization, or constant conjunction, concerning all crows. The idea that something's being a crow somehow *makes it* black arises only when this sentence is regarded as having the status of biological law, rather than that of a purely accidental generalization.

Likewise, the conclusion (7) can be regarded as ascribing a power or nomological regularity to King George's beliefs only if (7) is regarded as having the status of a law ascribing some special law-governed feature to George IV and his beliefs, rather than as an accidental constant conjunction. Now in deriving (7), we took (6) as our only premiss. Thus (7) may be regarded as stating some sort of law only if (6) may be.

Church remarks that, even though (6) is not certain, it is very likely. This observation may support a plausible view of (6) as some sort of psychological law concerning George IV and his beliefs. In this way, (7) would emerge as a law ascribing a nomological feature to King George's beliefs. Since no such law in fact obtains, and may even be falsified by the very case of Sir

Walter Scott and the author of *Waverley*, the meaningfulness of quantification into belief contexts, and therewith the theory of structured singular propositions, would be thereby discredited.

On the theory that I advocate, however, (6) is not only not very likely, as of some particular date during King George's acquaintance with Scott, it is very likely false.

It may seem as if denying (6) is tantamount to saying that George IV believed of some x that it is distinct from itself, and this seems a serious charge indeed. If an interest in the law of identity can hardly be attributed to the first gentleman of Europe, it is nothing short of blasphemy to attribute to him an interest in denying that law. In claiming that (6) is very likely false, as of some appropriate date, I mean no disrespect. Sentence (6) can easily be false even though King George is, of course, entirely rational—in fact, even if he were (what is beneath his dignity) a master of classical logic. If there was some time when George IV was acquainted with Scott and nevertheless believed after reading a *Waverley* novel that Scott was not the author, then (6) is false with respect to that time. If this be disputed, imagine instead that George IV confronted Scott at a book-signing ceremony, at which Scott truthfully proclaimed his authorship of *Waverley* but disguised himself in order to conceal his identity as Sir Walter Scott. Suppose the disguise succeeded in fooling even King George.[26] Let George IV say with conviction, pointing to the disguised author, 'He is not Sir Walter Scott'. In this case, (6) is decisively false. George IV is in the same unfortunate position as that of the ancient astronomer who believed of Venus that it is distinct from it.

Why, then, does Church claim that (6) is very likely? My conjecture is that Church confuses (6) with:

(6′) For every x, George IV does not believe that $(\lambda x')[x' \neq x'](x)$

or with:

(6″) For every x, George IV does not believe that x is self-distinct,

where the term 'self-distinct' is understood in accordance with the definition of the '*self*'-prefix given in Section III above. Both of these are indeed extremely likely—nay (I hasten to add), virtually certain. On the theory that I advocate, the pair of open sentences

$$x \neq x$$

and

$$(\lambda x')[x' \neq x'](x)$$

(or '*x* is distinct from *x*' and '*x* is self-distinct'), although logically equivalent, must be sharply distinguished as regards the propositions expressed under any particular assignment of a value to the variable '*x*'. Under the assignment of Scott to '*x*', the singular proposition contained in the first open sentence is believed by George IV in the book-signing example, the second is not. The extreme likelihood of (6′) and (6″) does not extend to (6).

Whereas sentences (6′) and (6″) are similar to, and easily confused with sentence (6), the former sentences do not concern King George's doxastic attitudes toward the propositions involved in sentence (6). They concern propositions of the form *x is self-distinct* (which ascribe the plainly impossible property of self-distinctness to particular individuals *x*) rather than propositions of the form *x is distinct from x* (which ascribe the relation of distinctness to reflexive pairs of individuals $\langle x, x\rangle$). Sentences (6′) and (6″) provide adequate explanation why George IV is disinclined to answer affirmatively when queried "Is Sir Walter self-distinct?", but the substitution of these sentences for Church's (6) does not show sufficient appreciation for the fact that King George is similarly disinclined when queried "Is Sir Walter distinct from Sir Walter?", or when any other similarly worded question is posed. These considerations give rise to a second potential confusion that could also lead one to conclude erroneously that (6) is true or at least very likely. By invoking the ternary *BEL* relation, something even closer to (6) may be assumed as at least very likely:

(6‴) For every *x*, if there is a *y* such that George IV is familiar with the proposition that $x \neq x$ by means of *y*, then there is a *y*′ such that George IV is familiar with the proposition that $x \neq x$ by means of *y*′ and not-*BEL*(George IV, that $x \neq x$, *y*′).

That is, either George IV is not familiar at all with the proposition that $x \neq x$ (in which case he does not believe it) or he withholds belief concerning whether $x \neq x$, either by disbelieving or by suspending judgement. (See the third condition on the *BEL* relation in Section V above.) Although (6‴) is not certain, it is very likely true as of the date in question, and this yields an explanation for King George's failure to assent to 'Sir Walter is distinct from Sir Walter'. But (by the first and second conditions on *BEL*) it does not follow that (6) itself is true or even likely.

It is entirely an empirical question whether (6) itself is true. There is no reason in advance of an actual investigation to suppose that (6) is even probably true.[27] By the same token, however, even if (6) is in fact very unlikely, it might well have been true throughout King George's lifetime. In some perfectly plausible alternative history of the world, it is true. If (6) were true, (7) would be as well. What then? Are we only contingently rescued from paradox in the actual world by the contingent falsity of (6)?

Even if (7) were true, it would not state a law ascribing some strange property to King George's beliefs. It would state a purely contingent constant conjunction concerning every pair of individuals x and y, an accidenal generalization that happens to be true not by virtue of some nomological feature of King George IV and his beliefs, but because—fortunately for King George—(6) happens to be true. No power to control the actual facts about x and y would be ascribed to King George's beliefs. If (6) were true (and Scott still had written *Waverley*), it would have to be true as well that King George does not believe that Scott is not the author of *Waverley*, and that George IV is not otherwise mistaken about the distinctness of any other pairs of identical objects of his acquaintance. The derivation of (7) from (6) would be sound, but it would no more constitute an unacceptable paradox than the so-called "paradoxes of material implication" constitute unacceptable paradoxes concerning 'if . . ., then'. In fact, since (7) employs the material 'if . . ., then', Church's paradox concerning quantification into belief contexts is a version of one of the "paradoxes of material implication".

VIII

What is the nature of the connection among the Richard–Soames problem, the puzzle of reflexives in propositional attitudes, and Church's "paradox" concerning quantification into belief attributions?

It is important to notice that, unlike the original puzzle generated by the Richard–Soames problem, neither the puzzle of reflexives in propositional attitudes nor Church's paradox makes essential use of existentially general beliefs, such as those ascribed in (1*c*), (2*c*), or (4*c*), or that denied in:

George IV does not believe that for some x, $x \neq x$.

Instead, the puzzle of reflexives in propositional attitudes and Church's paradox essentially employ beliefs whose formulation involves reflexive devices, such as the reflexive pronoun 'itself' and the '*self*'-prefix defined above. Conversely, the original puzzle, as constructed by means of sentences such as (1*b*), (2*b*), and (4*b*), makes no explicit use of beliefs whose formulations involve reflexive pronouns or other such devices. In lieu of such beliefs, the original puzzle employs beliefs whose formulations involve repeated occurrences of the same, or otherwise anaphorically related, bound variables or pronouns: the occurrences of '*x*' in (3*c*), the occurrences of 'it' in (2*c*) and (4*c*), the 'whom' and 'his' in (1*c*). In each case, these recurrences, or similarly related occurrences, are bound together from *within* the belief context. If I am correct, Church's "paradox" results, in part, from a confusion of a belief involving recurrences of the same variable bound together from *outside* the belief context with a belief involving a reflexive device. Nothing with the force of any of these puzzles is generated if we confine ourselves to beliefs involving recurrences of the same proper name, as in (1*b*), (2*b*), and (4*b*), or beliefs involving recurrences of the same variable or pronoun bound together from without, as ascribed in (3*d*) and (4*b*′) and denied in (6), and keep them sharply separated from beliefs involving reflexive devices or variables or pronouns bound together from within. On the theory formed from the conjunction of theses (*B*) and (*R*), sentences (1*b*), (2*b*), (4*b*), and (4*b*′) are all straightforwardly true. It appears likely, therefore, that the general phenomenon that gives rise to all three of these puzzles centres on some important element that is common to beliefs whose formulations involve reflexive devices and beliefs whose formulations involve recurrences of variables or pronouns bound together (from within any belief attribution), but absent from beliefs whose formulations involve recurrences of proper names or of free variables or pronouns (bound together from without the belief attribution).

Wherein is this common element of reflexivity? The question is significantly vague, and therefore difficult to answer. Some of the apparatus of *Frege's Puzzle*, however, points the way to a possible response.

In *Frege's Puzzle* the binding of a variable is regarded as involving the abstraction of a compound monadic predicate from an open sentence. Thus '($\exists x$)('Hesperus' refers to x and 'Phosphorus' refers to x)' is seen on the model of 'Something is such that 'Hesperus' refers to it and 'Phosphorus' refers to it', and '($\exists x$)(x outweighs x)' is seen on the model of 'Something is such that it outweighs it', where in each case the initial word 'something' is a second order predicate and the remainder of the sentence is the abstracted compound monadic predicate to which 'something' is attached. In fact, the abstracting of a predicate from an open sentence of formal logic using Church's 'λ'-operator might be understood on the model of transforming an "open" sentence such as 'I love *it* and *it* loves me' (with both occurrences of '*it*' functioning as "freely" as a free variable of formal logic) into the corresponding closed monadic predicate 'is such that I love it and it loves me'.

Compound monadic predicates formed by variable-binding (or pronoun-binding) abstraction from open sentences are treated in *Frege's Puzzle* as yielding an exception to the general rule that the contribution to information content made by (i.e. the "information value" of) a compound expression is a complex entity made up of the contributions of the components. Instead such compound predicates are taken as contributing a semantically associated temporally indexed property, taken as a unit. (See note 5.) Thus, the (closed) abstracted predicate 'is an object x such that 'Hesperus' refers to x and 'Phosphorus' refers to x' is regarded as contributing, with respect to a time t, simply the property of being referred to at t by both 'Hesperus' and 'Phosphorus', and the (closed) abstracted predicate 'is an object x such that x outweighs x' is regarded as contributing, with respect to t, the property of outweighing oneself at t. The proposition contained, with respect to t, by 'Something is such that it outweighs it' (or 'Something is an object x such that x outweighs x') is taken as being composed of this latter property together with the contribution made by 'something' (to wit, the property of being a nonempty class at t).

The properties of being referred to at t by 'Hesperus' and also by 'Phosphorus' and of outweighing onself at t contain the element of reflexivity that also arises when using the '*self-*'prefix, defined in Section III above by means of the binding of a recurring variable. The dyadic-predicate-operator defined in Section VI above in connection with the question of the meanings of reflexive pronouns also involves the binding of a recurring variable, and thereby also involves this element of reflexivity. Some such aspect of the binding of recurring variables and pronouns seems to provide the link among the Richard–Soames problem, the puzzle of reflexives in propositional attitudes, and Church's paradox concerning quantification into belief contexts.

Notes

* Many of the ideas in this paper were first urged by me in correspondence with David Kaplan, Mark Richard, and Scott Soames in February 1984. There was also a discussion of some of these issues with Joseph Almog, Kaplan, and Soames, and some later correspondence with Alonzo Church. Although there was not the time before submission to receive reactions or comments on the present paper, it has benefited from these earlier exchanges.

1. M. Richard, 'Direct Reference and Ascriptions of Belief', *Journal of Philosophical Logic* 12 (1983), 425–52; also in this volume.
2. 'Lost innocence', *Linguistics and Philosophy* 8 (1985), 59–71.
3. MIT Press, 1983.
4. MIT Press, 1986.
5. The idea of indexing, or relativizing, the notion of information content to times (independently of contexts) is due to M. Richard, 'Tense, Propositions, and Meanings', *Philosophical Studies* 41 (1982), 337–51. The idea that the contribution made by a predicate to information content is something like a temporally indexed attribute is defended in *Frege's Puzzle* and stems from Richard's idea of indexing information content to times.
6. B. Russell, 'The Philosophy of Logical Atomism', in R. C. Marsh (ed.), *Logic and Knowledge* (George Allen and Unwin: London, 1956), 177–281; also pp. 35–155 in Russell's *The Philosophy of Logical Atomism*, ed. D. Pears (Open Court: La Salle, 1985).
7. S. Kripke, 'A Puzzle about Belief' in A. Margalit (ed.) *Meaning and Use*, (D. Reidel: Dordrecht, 1979), 239–75; also in this volume.
8. See K. Donnellan 'Proper Names and Identifying Descriptions', in D. Davidson and G. Harman (eds.), *Semantics of Natural Language* (D. Reidel: Dordrecht, 1972), 356–79; S. Kripke *Naming and Necessity* (Harvard University Press and Basil Blackwell, 1972, 1980); also in D. Davidson and G. Harman (eds.), *Semantics of Natural Language*, (D. Reidel: Dordrecht, 1972), 253–355, 763–9; and J. Perry 'The Problem of the Essential Indexical', *Nous* 13 (1979), 3–21; also in this volume. For a summary of the major difficulties with the views of Frege and Russell, see N. Salmon, *Reference and Essence* (Princeton University Press and Basil Blackwell, 1981), chapter 1. Further problems with the Frege–Russell view in connection with propositional attitudes are discussed in *Frege's Puzzle*, ch. 9 and *passim*.
9. 'Direct Reference and Ascriptions of Belief'; pp. 440–2 and *passim*. He constructs (pp. 444–5) a semantics for belief attributions that conflicts with thesis (*B*).
10. 'Lost Innocence', p. 62.
11. Richard 'Direct Reference and Ascriptions of Belief' is one exception.

12. The reason for the phrase 'by and large' is that there are important classes of exceptions to the general rule. Certain nonextensional operators, such as quotation marks, create contexts in which compound expressions contribute themselves as units to the information content of sentences in which the quotation occurs, and other nonextensional operators, such as temporal operators, create contexts in which some compound expressions contribute complexes other than their customary contribution to information content (see n. 5, above). In addition, we shall see below that a compound predicate formed by abstraction from an open sentence is regarded as contributing something like an attribute, taken as unit, rather than a complex made up of the typical contributions of the compound's components.
13. The '*self*'-prefix defined here may not correspond exactly to that of ordinary English. In English, the term 'self-cleaning' may apply, with respect to a time *t*, to an object even if that object is not cleaning itself *at t* (say, because it is unplugged or switched off for the moment), as long as the object is the *sort* of thing at *t* that cleans itself at appropriate times. Similarly, someone is self-indulgent at *t* if and only if he or she is the sort of person at *t* that has at some appropriate times the feature of indulging oneself, even if he or she is not doing so at *t*. The '*self*'-prefix defined here is such that ⌜*self-R*⌝ applies to an object with respect to *t* if and only if the object bears *R* to itself at *t*.
14. The unrestricted rule of exportation has been shown invalid, or at least highly suspect, by the fallacious inference 'The shortest spy exists and Ralph believes: that the shortest spy is a spy; therefore Ralph believes of the shortest spy: that he or she is a spy'. From the conclusion of this inference one may validly infer 'There is someone whom Ralph believes to be a spy', which intuitively does not follow from the initial premiss. This instance of exportation fails because the exported term, 'the shortest spy', is a definite description. The theory formed from the conjunction of theses (*B*) and (*R*) requires the validity of exportation with respect to belief attributions, provided the rule is restricted to proper names, demonstratives, or other simple singular terms. Hence, the theory must accept the inference from (3*b*) to (3*b*′) (assuming the tacit premiss ⌜$(\exists x)\ [x = a]$⌝). Similarly, the theory is committed to the validity of importation, inferring ⌜*c* believes: that ϕ_a⌝ from ⌜*c* believes of *a*: that ϕ_{it}⌝, under the same restriction on *a*.
15. 'A Remark concerning Quine's Paradox about Modality', Spanish translation in *Analisis Filosofico* 2. 1–2 (May–November 1982), 25–34; in English in this volume.
16. He compares his result to the derivation in standard quantified modal logic of the contrapositive of the necessity of identity: If any *x* and *y* can be distinct, they are. He likewise cites the derivability of this principle (which he calls 'a variant of Murphy's Law') as providing a reason for rejecting the meaningfulness of quantification into modal contexts.

    Church seems to allow that on the theory of structured singular propositions as potential objects of belief (which he calls 'the principle of transparency of belief' and which he regards as a doubtful theory), a power to control the actual facts about *x* and *y* with one's beliefs would not be surprising and could be explained. Unfortunately, he does not provide the alleged explanation. I am unsure what he has in mind with his remarks in this connection. Speaking as one who is deeply committed to the theory in question, I would find such a power surprising in the extreme and utterly inexplicable. I see the problem as one of how to reconcile the derivation of (7) with the obvious fact that no such power exists (and the fact that (7) may even be false in the case of Sir Walter Scott and the author of *Waverley*).

17. That is, neither Lois nor the astronomer knows (in the example) that she or he believes some proposition together with its denial. On the other hand, Lois does know that she believes that she will inform Superman of his danger with her note and will not inform Superman of his danger with her note, and the astronomer does know that he believes that 'Hesperus' refers to Venus and 'Phosphorus' refers to Venus and 'Hesperus' and 'Phosphorus' do not both refer to Venus. Furthermore, each presumably knows that these propositions are contradictory—though neither knows that she or he believes a contradictory proposition. Sorting these matters out is a delicate task made extremely difficult by relying on the term 'believes' without the use of some expression for the full ternary *BEL* relation. Cf. *Frege's Puzzle*, ch. 8.

18. Soames has offered an account not unlike this one in response to my urging on him that, as Richard originally presented the problem, it poses a serious difficulty for the theory of structured singular propositions as well as for the set-of-circumstances theories. See 'Lost Innocence', p. 69 n. 12. The notion of a "belief state" invoked there (which seems to have been derived from 'The Problem of the Essential Indexical' and *Situations and Attitudes*) plays a role analogous to the third relata of the *BEL* relation in my account, the *ways* in which one may be familiar with, or take, a proposition. See *Frege's Puzzle*, ch. 9 n. 1, for some brief remarks comparing the third relata of the *BEL* relation (proposition guises, or modes of acquaintance with propositions) with Perry's notion of a belief state.

    In response to the Richard–Soames problem, Barwise and Perry seem to have abandoned the idea that the information content ("interpretation") of a declarative sentence, with respect to a context, is the set of situations (or type of situation) with respect to which the sentence is true (with 'situation' understood in such a way that any situation with respct to which it is true that 'Hesperus' and 'Phosphorus' both refer to Venus is one with respect to which it is also true that there is something referred to by both 'Hesperus' and 'Phosphorus'). In fact, Barwise and Perry seem to have moved significantly in the direction of structured singular propositions and an account of the Richard–Soames problem similar (in certain respects) to the one advanced here and to the one advanced by Soames (see 'Shifting Situations and Shaken Attitudes', *Linguistics and Philosophy* 8 (1985), 105–61 (also available as the Stanford University Center for the Study of Language and Information Report No. CSLI-84-13), pp. 153–8, esp. pp. 156–7). If so, this move constitutes an important concession to Soames. However, Barwise and Perry (if I understand them correctly) couple this move with the surprising claim (p. 158) that there is a significant sense in which the information content of "Something or other is referred to by both 'Hesperus' and 'Phosphorus' " is not a consequence of that of "Venus exists and 'Hesperus' refers to it and 'Phosphorus' refers to it". If this is to be understood as a claim about the logic of the information contents of these sentences, surely the claim must be rejected, and the doctrine of structured singular propositions as the information contents ("interpretations") of sentences and the objects of belief, coupled with classical logic, is much to be preferred over the newer theory of Barwise and Perry.

19. 'A Proposed Solution to a Puzzle about Belief', in P. French, T. Uehling, and H. Wettstein (eds.) *Midwest Studies in Philosophy, vi. The Foundations of*

Analytic Philosophy, (University of Minnesota Press, 1981), 501–10, esp. 503–6, and 'Rationality and Believing the Impossible', *Journal of Philosophy* 80, 6 (June 1983), 321–38.

20. See 'A Proposed Solution to a Puzzle about Belief', p. 505, and 'Rationality and Believing the Impossible', p. 330. Marcus's argument focuses on the special case where *R* is a predicate for numerical distinctness. She writes: "If I had believed that Tully is not identical with Cicero, I would have been believing that something is not the same as itself and I surely did not believe that, a blatant impossibility, so I was mistaken in claiming to *have* the belief [that Tully is not Cicero]", and, "[believing that London is different from Londres] would be tantamount to believing that something was not the same as itself, and surely I could never believe *that.* So my belief claim [my claim that I believed that London is not Londres] was mistaken . . .". These arguments evidently rely on the premiss that if (3*a*) then (3*c*) (or perhaps on the premiss that if (3*a*) then [the existential generalization on *a* of (3*e*′)], i.e. if *c* believes that *a* and *b* are distinct, then *c* believes *of* something that it is distinct from itself).
21. *Structured Meanings: The Semantics of Propositional Attitudes* (MIT Press, 1985), p. 23.
22. The argument presented thus far has been emphasized by D. Wiggins in a number of writings. See e.g. 'Identity, Necessity and Physicalism', in S. Körner (ed.), *Philosophy of Logic* (University of California Press, 1976), 96–132, 159–82, esp. 164–6; and 'Frege's Problem of the Morning Star and the Evening Star', in M. Schirn (ed.), *Studies on Frege*, ii. *Logic and the Philosophy of Language* (Bad Canstatt: Stuttgart, 1976), 221–55, esp. 230–1.

    Wiggins credits the argument to Peter Geach, and claims to have extracted the argument from Geach *Reference and Generality* (Cornell University Press, 1962), p. 132. This, however, is a serious misinterpretation of Geach, whose view is precisely the denial of Wiggins's view that sentences of the form ⌜*a* bears *R* to *a*⌝ and ⌜*a* bears *R* to itself⌝ contain different information, or express different propositions. See e.g. Geach 'Logical Procedures and the Identity of Expressions', in id. *Logic Matters* (University of California Press, 1972), 108–15, esp. 112–13. If I read Geach correctly, his view is that a sentence such as 'Hesperus outweighs Hesperus' ascribes to Venus the reflexive property of weighing more than oneself, as does the sentence 'Hesperus outweighs itself', rather than the simple relational property of weighing more than Venus. (Cf. the treatment of the contents of sentences with recurring expressions in H. Putnam, 'Synonymy, and the Analysis of Belief Sentences', *Analysis* 14, 5 (April 1954), 114–22; also in this volume. See also *Frege's Puzzle*, pp. 164–5 n. 4.) The argument in *Reference and Generality* is intended to show not that 'Marx contradicts himself' differs in information content from 'Marx contradicts Marx' (which Geach rejects), but that the 'himself' in 'Marx contradicts himself' is not a singular term referring to Marx. The argument for this conclusion (which Wiggins presumably also believes) is part of a defence of Geach's general view that pronouns occurring with antecedents are typically not referring singular terms. (I disagree with Geach both concerning the semantic analysis of sentences such as 'Hesperus outweighs Hesperus' and 'Marx contradicts Marx', as does Wiggins, and concerning typical (nonreflexive) pronouns with antecedents, though the latter issue is not germane to the topic of the present discussion.)

    The same (or very nearly the same) misinterpretation of Geach's argument in

Reference and Generality occurs in G. Evans 'Pronouns, Quantifiers and Relative Clauses (I)', in M. Platts (ed.), *Reference, Truth and Reality*, (Routledge and Kegan Paul: London, 1980), 255–317, esp. 267–8, although Evans admits that the view he attributes to Geach has been unambiguously denied by Geach in a number of places. (Oddly, Evans cites references to, and even quotes, writings in which Geach clearly denies the view that Evans attributes to him.) It might be said that in accusing Geach of misinterpreting Geach, Evans takes on the very property he attributes to Geach—although Evans does not misinterpret himself.

23. The response to the Richard–Soames problem in 'Shifting Situations and Shaken Attitudes' suggests that Barwise and Perry might similarly respond to the puzzle of reflexives in propositional attitudes by claiming that 'Hesperus outweighs Hesperus' and 'Hesperus is self-outweighing' differ in information content (have different "interpretations"). See n. 18 above. Such a move would constitute a repudiation of the idea, fundamental to *Situations and Attitudes*, that the information content ("interpretation") of a sentence is the set of situations (or type of situation) with respect to which the sentence is true—with the term 'situation' understood in such a way that any situation with respect to which it is true that Venus outweighs Venus is one with respect to which it is also true that Venus is self-outweighing, and vice versa. Any attempt to modify their view to accommodate the fact that 'Hesperus outweighs Hesperus' and 'Hesperus is self-outweighing' differ in information content would clearly constitute a concession to the structured-singular-proposition sort of theory advocated in *Frege's Puzzle*.

    However, this move might be coupled with the claim that there is a significant sense in which the second information content ("interpretation") is no consequence of the first. They might even claim that there is a significant sense in which these information contents are independent and neither implies the other. Here again, if either of these claims is to be understood as concerning the logic of the information contents of 'Hesperus outweighs Hesperus' and 'Hesperus is self-outweighing', they must surely be rejected. Insofar as the newer theory of Barwise and Perry includes one or both of these claims, the doctrine of structured singular propositions coupled with the denial of each of these claims (and with classical logic) is much the preferable theory.

24. This part of the argument is intended as a rejoinder to Evans's response in 'Pronouns, Quantifiers and Relative Clauses', p. 268.

25. Cf. *Reference and Generality*, pp. 136–7. In the theoretical apparatus of *Frege's Puzzle*, the contribution to information content made by (i.e. the "information value" of) the displayed expression is the *operation* of assigning to any class K of ordered pairs of individuals the class of individuals i such that the reflexive pair $\langle i, i\rangle \in K$.

26. As a matter of historical fact, Scott did conceal his authorship of *Waverley*, and George IV did wish to know whether Scott indeed wrote *Waverley*. Hence Russell's clever example.

27. This contrasts sharply with the analogous principle involved in the standard proof in quantified modal logic of the necessity of identity: For every x, it is not possible that $x \neq x$. This is a logical truth, and therefore an *a priori* certainty. Unlike (7), the necessity of identity (or equivalently, Murphy's Law of Modality) is a genuine law (in this case, a law of logic).

NOTES ON THE CONTRIBUTORS

ALONZO CHURCH is Flint Professor of Philosophy and Mathematics at the University of California, Los Angeles. A member of the National Academy of Sciences, he is the author of *Introduction to Mathematical Logic*, as well as numerous articles in mathematical logic and the philosophy of language. His contributions to mathematical logic include the 'λ'-calculus, *Church's Thesis*, and *Church's Theorem*. He has done extensive work defending and developing Frege's philosophico-semantic theory.

GOTTLOB FREGE (1848–1925) taught mathematics at the University of Jena. A seminal figure in the philosophy of mathematics and the philosophy of language, he was the principal founder of modern mathematical logic. His works include *Begriffsschrift* (1879), *Grundlagen der Arithmetik* (1884), and *Grundgesetze der Arithmetik* (1893). His classic 'Uber Sinn und Bedeutung' (1892) introduced his celebrated theory of sense and denotation, which helped lay the foundation for contemporary philosophy of language.

DAVID KAPLAN is Professor of Philosophy at the University of California, Los Angeles. He has written both influential and uninfluential articles in philosophical logic and the philosophy of language, including 'Quantifying In', 'Demonstratives', and 'Opacity'. He delivered Oxford University's John Locke Lectures in 1980. He is honoured in a forthcoming festschrift *Themes from Kaplan*.

SAUL KRIPKE is McCosh Professor of Philosophy at Princeton University. A pivotal figure in the fields of modal and intuitionistic logic, he is the author of several influential works in logic and philosophy, including 'A Completeness Theorem in Modal Logic', 'Semantical Considerations on Modal Logic', *Naming and Necessity*, 'Outline of a Theory of Truth', and *Wittgenstein on Rules and Private Language*. In logic, he is also known for his theory of transfinite recursion on admissible ordinals. He delivered the John Locke Lectures in 1973.

JOHN PERRY is Henry Waldgrave Stuart Professor of Philosophy at Stanford University. He is the author of several influential articles in the philosophy of mind and the philosophy of language, and is co-author, with Jon Barwise, of *Situations and Attitudes*.

HILARY PUTNAM is Walter Beverly Pearson Professor of Mathematical Logic at Harvard University. He is the author of numerous influential articles, many of which are collected in his three-volume *Philosophical Papers*. His books include *Meaning and the Moral Sciences* (based on the John Locke Lectures), *Reason, Truth and History*, and *The Many Faces of Realism* (the Carus Lectures).

MARK RICHARD is Assistant Professor of Philosophy at Tufts University. He has written on philosophical logic and the philosophy of language.

BERTRAND RUSSELL (1872–1969) was one of the great figures of twentieth century philosophy. A prolific writer, his philosophical inventions and discoveries include Russell's Paradox, the Theory of Types, the Theory of Descriptions, and the philosophy of logical atomism. He was awarded the Nobel Prize for Literature in 1950.

NATHAN SALMON is Professor of Philosophy at the University of California, Santa Barbara. He is the author of a number of articles in metaphysics and the philosophy of language, and two books, *Reference and Essence* and *Frege's Puzzle*.

SCOTT SOAMES is Associate Professor of Philosophy at Princeton University. He is the author of several articles in linguistics and the philosophy of language, and is co-author, with David Perlmutter, of *Syntactic Argumentation and the Structure of English*.

SELECT BIBLIOGRAPHY
(Not including material in this volume)

ANDERSON, C. A. 'General Intensional Logic', in D. Gabbay and F. Guenthner (eds.), *Handbook of Philosophical Logic*, ii: *Extensions of Classical Logic* (Dordrecht: D. Reidel, 1984), 355–85.

BARWISE J. and PERRY, J., *Situations and Attitudes* (MIT Press/Bradford Books, 1983).

BOËR S. and LYCAN, W., 'Who Me?', *The Philosophical Review* 89 (1980), 427–66.

BURGE, T., 'Belief and Synonymy', *Journal of Philosophy* 75 (1978), 119–38.

CARNAP, R., *Meaning and Necessity* (University of Chicago Press, 1947).

CARTWRIGHT R., 'Propositions', in R. Butler (ed.), *Analytic Philosophy*, 1st ser. (Oxford: Basil Blackwell, 1966), 81–103.

CHURCH, A., Review of Carnap, *Introduction to Semantics*, in *The Philosophical Review* 52 (1943), 298–304.

—— 'On Carnap's Analysis of Statements of Assertion and Belief', *Analysis* 10 (1950), 97–9; also in L. Linsky (ed.), *Reference and Modality* (Oxford University Press, 1971), 168–70.

—— 'A Formulation of the Logic of Sense and Denotation', in P. Henle, H. Kallen, and S. Langer (eds.), *Structure, Method and Meaning: Essays in Honor of Henry M. Sheffer* (New York: Liberal Arts Press, 1951), 3–24.

—— *Introduction to Mathematical Logic*, vol. i (Princeton University Press, 1956).

—— 'Outline of a Revised Formulation of the Logic of Sense and Denotation', Part I: *Noûs* 7 (March 1973), 24–33; Part II: *Noûs* 8 (May 1973), 135–56.

DONNELLAN, K., 'Reference and Definite Descriptions', *The Philosophical Review* 75 (1966), 281–304; also in S. Schwartz (ed.), *Naming, Necessity, and Natural Kinds* (Cornell University Press, 1977), 42–65.

—— 'Proper Names and Identifying Descriptions', in D. Davidson and G. Harman (eds.), *Semantics of Natural Language* (Dordrecht: D. Reidel, 1972), 356–79.

EVANS, G., *Varieties of Reference* (Oxford University Press, 1982).

FREGE, G., *'Uber Sinn und Bedeutung', Zeitschrift für Philosophie und Philosophische Kritik* 100 (1893), 25–50; English translation in B. McGuinness (ed.), *Collected Papers on Mathematics, Logic, and Philosophy*, trans. by M. Black, V. H. Dudman, P. Geach, H. Kaal, E.-H. W. Kluge, B. McGuinness, and R. H. Stoothoff (Oxford: Basil Blackwell, 1984), 157–77; also in P. Geach and M. Black, *Translations from the Philosophical Writings of Gottlob Frege* (Oxford: Basil Blackwell, 1952), 56–78.

GRICE, H. P. 'Logic and Conversation', in D. Davidson and G. Harman (eds.), *The Logic of Grammar* (Encino: Dickenson, 1975), 64–75.

KAPLAN, D., 'Quantifying In', in D. Davidson and G. Harman (eds.), *Words and Objections: Essays on the Work of W. V. Quine* (Dordrecht: D. Reidel, 1969), 206–42; also in L. Linsky (ed.), *Reference and Modality* (Oxford University Press, 1971), 112–44.

—— 'Bob and Carol and Ted and Alice', in J. Hintikka, J. Moravcsik, and P. Suppes (eds.), *Approaches to Natural Language* (Dordrecht: D. Reidel, 1973), 490–518.

—— 'How to Russell a Frege–Church', *Journal of Philosophy* 72 (1975), 716–29; also in M. Loux (ed.), *The Possible and the Actual* (Cornell University Press, 1979), 210–24.

—— 'Dthat', in P. Cole (ed.), *Syntax and Semantics*, 9: *Pragmatics* (New York: Academic Press, 1978), 221–43; also in French, Uehling and Wettstein (eds.), *Contemporary Perspectives in the Philosophy of Language* (University of Minnesota Press, 1979), 383–400.

—— 'Opacity', in L. E. Hahn and P. A. Schilpp (eds.), *The Philosophy of W. V. Quine* (La Salle, Ill.: Open Court, 1986), 229–88.

—— 'Demonstratives', in J. Almog, J. Perry, and H. Wettstein (eds.), *Themes from Kaplan* (Oxford University Press, 1988).

KRIPKE, S., 'Identity and Necessity', in M. Munitz (ed.), *Identity and Individuation* (New York University Press, 1971), 135–64; also in S. Schwartz (ed.), *Naming, Necessity, and Natural Kinds* (Cornell University Press, 1977), 66–101.

—— *Naming and Necessity* (Harvard University Press and Basil Blackwell, 1972, 1980); also in D. Davidson and G. Harman (eds.), *Semantics of Natural Language* (Dordrecht: D. Reidel, 1972), 253–355, 763–69.

—— 'Speaker's Reference and Semantic Reference', in P. French, T. Uehling, and H. Wettstein (eds.), *Contemporary Perspectives in the Philosophy of Language* (University of Minnesota Press, 1979), 6–27.

LEWIS, D., 'Attitudes *De Dicto* and *De Se*', *The Philosophical Review* 88 (1979), 513–43; also in Lewis's *Philosophical Papers*, vol. i (Oxford University Press, 1983), 133–59.

MARCUS, R. B. 'Rationality and Believing the Impossible', *Journal of Philosophy* 80 (1983), 321–38.

MATES, B., 'Synonymity', *University of California Publications in Philosophy* 25 (1950); also in L. Linsky (ed.), *Semantics and the Philosophy of Language* (University of Illinois Press,1952), 111–36.

MILL, J. S., 'Of Names', in Book I, Chapter II of *A System of Logic* (New York: Harper and Brothers, 1893), 29–44.

PERRY, J., 'Frege on Demonstratives', *The Philosophical Review* 86 (1977), 474–97.

PUTNAM, H., 'Is Semantics Possible?', in H. E. Keifer and M. Munitz

(eds.), *Language, Belief, and Metaphysics* (State University of New York Press, 1970), 50–63; also in S. Schwartz (ed.), *Naming, Necessity, and Natural Kinds* (Cornell University Press, 1977), 102–18.
—— 'The Meaning of "Meaning"', in K. Gunderson (ed.), *Minnesota Studies in the Philosophy of Science*, vii: *Language, Mind, and Knowledge* (University of Minnesota Press, 1975); also in Putnam's *Philosophical Papers*, ii: *Mind, Language, and Reality* (Cambridge University Press, 1975), 215–71.
QUINE, W. V. O., 'Reference and Modality', in Quine, *From a Logical Point of View* (New York: Harper and Row, 1953), pp. 139–59; also in L. Linsky (ed.), *Reference and Modality* (Oxford University Press, 1971), 17–34.
—— 'Quantifiers and Propositional Attitudes', *Journal of Philosophy* 53 (1956), 177–87; also in Quine, *The Ways of Paradox* (New York: Random House, 1966), 183–94; and L. Linsky (ed.), *Reference and Modality* (Oxford University Press, 1971), 101–11.
RICHARD, M., 'Attitude Ascriptions, Semantic Theory, and Pragmatic Evidence', *Proceedings of the Aristotelian Society* (1987), 243–62.
—— 'Quantification and Leibniz's Law', *The Philosophical Review* 96 (1987), 555–78.
RUSSELL, B., 'On Denoting', *Mind* 14 (1905), 479–93; also in Russell's *Logic and Knowledge*, ed. R. C. Marsh (London: George Allen and Unwin, 1956), 41–56.
—— 'The Philosophy of Logical Atomism', in Russell, *Logic and Knowledge*, ed. Marsh R. C. (London: George Allen and Unwin, 1956), 177–281; also in Russell, *The Philosophy of Logical Atomism*, ed. D. Pears (La Salle: Open Court, 1985), 35–155.
—— 'Mr. Strawson on Referring', *Mind* 66 (1957), 385–9.
SALMON, N., *Reference and Essence* (Princeton University Press and Basil Blackwell, 1981).
—— *Frege's Puzzle* (MIT Press/Bradford Books, 1986).
—— 'Illogical Belief', in J. Tomberlin (ed.), *Philosophical Perspectives*, 3: *Philosophy of Mind and Action Theory* (Atascadero: Ridgeview, forthcoming 1989).
—— 'Reference and Information Content: Names and Descriptions', in D. Gabbay and F. Guenthner (eds.), *Handbook of Philosophical Logic*, iv: *Topics in the Philosophy of Language* (Dordrecht: D. Reidel, forthcoming).
SMULLYAN, A., Review of Quine, 'The Problem of Interpreting Modal Logic', in *The Journal of Symbolic Logic* 12 (1947), 139–41.
SOAMES, S., 'Lost Innocence', *Linguistics and Philosophy* 8 (1985), 59–71.
—— 'Substitutivity', in J. J. Thomson (ed.), *On Being and Saying: Essays for Richard Cartwright* (MIT Press, 1987), 99–132.
STALNAKER, R., *Inquiry* (MIT Press/Bradford Books, 1984).
STRAWSON, P. E., 'On Referring', *Mind* 59 (1950), 320–44.

INDEX OF NAMES